PLUMBING & HEATING

PLUMBING & HEATING

Created and editorially produced for Petersen Publishing Company by Allen D. Bragdon Publishers, Inc., Home Guides Division

STAFF FOR THIS VOLUME:

Editorial Director	Allen D. Bragdon
Managing Editor	Michael Donner
Art Director	John B. Miller
Assistant Art Director	Lillian Nahmias
Text Editor	Donal Dinwiddie
Copy Editors	Joe Fernandez, Hannah Selby
Contributing Artists	Pat Lee, Chuck Pitaro, Clara Rosenbaum, Jerry Zimmerman
Contributing Photographers	Jack Abraham, Donal Dinwiddie, Jay Hedden, Jayne Lathrop, Michael Mertz
Cover design by	"For Art Sake," Inc.
Text originated by	Doug Day (Plumbing), Joseph H. Foley (Heating)

Doug Day is the son of longtime DIY writer Richard Day. He has had do-it-yourself articles published in POPULAR SCIENCE and HOW TO magazine. He works in maintenance for the California State Department of Parks and Recreation.

Joseph H. Foley is Senior Publications Engineer for Sperry Rand Corporation. He has 30 years experience in writing instructions to make technical procedures clear and complete for people who know nothing about the subject.

ACKNOWLEDGEMENTS

The Editors wish to thank the following individuals and firms for their help in the preparation of this book: American Iron and Steel Institute; American Standard, Inc.; Bernzomatic Corp.; Bituminous Pipe Institute; Black and Decker Mfg. Co.; Carrier Corporation; Cast Iron Soil Pipe Institute; Chicago Pneumatic Tool Company; Cinch-Pipe System; Clow, Waste Treatment Division; Committee of Steel Pipe Producers; Copper Development Association, Inc.; Cromaglass Corp; Culligan, Inc.; Deeprock Manufacturing Company; Delta Faucet Company; General Electric Company; Genova, Inc.; Lennox Industries, Inc.; Montgomery Ward and Co.; Mortell Company; Peerless Faucet Company; Ridge Tool Co.; Sears, Roebuck and Co.; the Stanley Works; Sunstream, a division of Grumman Houston Corporation; U.S. Department of Agriculture; Water Pollution Control Federation; Water Well Journal, Anita Stanley; Westinghouse Electric Corp.

Petersen Publishing Company

R. E. Petersen/Chairman of the Board
F. R. Waingrow/President
Alan C. Hahn/Director, Market Development
James L. Krenek/Director, Purchasing
Louis Abbott/Production Manager
Erwin M. Rosen/Executive Editor,
Specialty Publications Division

Created by Allen D. Bragdon Publishers, Inc. Copyright © 1977 by Petersen Publishing Company, 8490 Sunset Blvd., Los Angeles, Calif. 90069. Phone (213) 657-5100. All rights reserved. No part of this book may be reproduced without written permission. Printed in U.S.A.

Library of Congress Catalog Card No. 77-076142
Paperbound Edition:
0-8227-8001-1
Hardcover Edition:
0-8227-8013-5

The text and illustrations in this book have been carefully prepared and fully checked to assure safe and accurate procedures. However, neither the editors nor the publisher can warrant or guarantee that no inaccuracies have inadvertently been included. Similarly, no warranty or guarantee can be made against incorrect interpretation of the instructions by the user. The editors and publisher are not liable for damage or injury resulting from the use or misuse of information contained in this book.

Table of Contents

1.	BASICS	4
	Getting started	4
	Home-buyer's checklist	5
	Water-supply system	6
	Parts of the system	6
	How an air chamber works	7
	Drain-waste-vent system	8
	Parts of the system	8
	Drain-waste-vent pipe sizes	9
	Fixture and appliance hookup	10
	Plumbing tools	12
	How to use basic tools	12
	Tips for general plumbing tools	14
	Tools for plastic pipe	15
2.	FAUCETS	16
	Faucet repairs	16
	Washer-type faucets	16
	Replacing a washer	17
	Fixing valve seats	18
	Disc faucet repair	19
	Lever faucet repair	19
	Washerless faucets	19
	Troubleshooting faucet leaks	20
	Fixing aerators	20
	Lever-type washerless faucets	21
	About valves	22
	Cross connections and vacuum-breaker valves	23
3.	SINKS AND BATHS	24
	Installing faucets	24
	Modernizing fixtures	24
	Replacing faucets	25
	Installing traps	26
	Replacing a sink sprayer	27
	Upgrading a shower head	27
	Installing fixture shutoffs	28
	Caulking	29
4.	TOILETS	30
	Unveiled: the mystery of the flush toilet	30
	Troubleshooting toilet tanks	31
	Flush valve parts	33
	Tankless flush valves	33
	Troubleshooting tankless flush valves	33
	Replacing a toilet	34
	Removing the old toilet	34
	Installing the new toilet	35
5.	PROBLEM SOLVING	36
	Unclogging drains	36
	Freeing sink or lavatory drains	36
	Freeing tub drains	37
	Freeing floor drains	37
	Freeing toilet drains	38
	Freeing branch and main drains	38
	Solving frequent clogging	39
	Fixing leaks	40
	Troubleshooting leaks	41
	Curing condensation	42
	Preventing freezing	42
	Thawing pipes	42
	Curing water hammer	43
6.	PIPING	44
	Vinyl water-supply piping	44
	Threaded water-supply piping	46
	Joining sweat-soldered copper tubing	47
	Running water-supply pipes	48
	Cast-iron DWV piping	49
	Plastic DWV piping	50
	Solvent welding	51
	Running a vent stack	51
	Sweat-soldered copper DWV	52
	Running DWV pipes	53
	Sizing DWV runs	54
	Sewer pipe is different	55
	Comparison of popular sewer pipes	55
	Adapting pipes	56
	Problems with low flow	57
7.	APPLIANCE REPAIR	58
	Clothes washers	58
	Troubleshooting an automatic clothes washer	58
	Dishwasher repairs	59
	Troubleshooting a dishwasher	59
	Sump pumps	60
	Replacing a sump pump	61
	Dehumidifiers	62
	Caring for your water heater	63
	Troubleshooting water heaters	63
	Water treatment	64
	How a water softener works	64
	Water-softener connections	64
	Arrangement of water-treatment units	65
8.	PRIVATE SYSTEMS	66
	The hydrologic cycle	66
	Water supply	66
	Drilling a well	67
	Connecting a pump	67
	How a pressure tank works	67
	Lowering a submersible pump	68
	Troubleshooting well pumps	68
	Hand-dug well	69
	Comparison of well types	69
	Sewage treatment	70
	Parts of a septic system	70
	Building a sewer	71
	Installing a tank	71
	Percolation table	71
	Tank-size table	71
	Building a seepage field	72
	Single-family sewage-treatment plant	73
9.	HEATING	74
	Heat moves in three ways	74
	Elements of a heating system	75
	Humidity	75
	Heat producers	76
	Oil burners	76
	How to maintain an oil burner	77
	How to restart an oil burner	77
	Draft control	77
	Gas burners	78
	Electric heating	79
	Heat exchangers, distributors, and conduits	80
	Routine maintenance	81
	Draining the system	81
	Draining the expansion tank	81
	Flushing and refilling	81
	Bleeding the line	81
	Forced hot-air systems	82
	Maintenance and adjustment	82
	Adjusting blower speed	83
	Pulley alignment	83
	Steam-heating systems	84
	Routine maintenance	84
	Fireplaces	85
	Solar heating	86
	Heat pumps	88
	Where heat pumps are most efficient	89
	How a heat pump works	89
10.	PROJECTS	90
	Cold-weather house shutdown	90
	Shutdown checklist	90
	Building a dry well	91
	• GLOSSARY	92
	• INDEX	94

1. BASICS

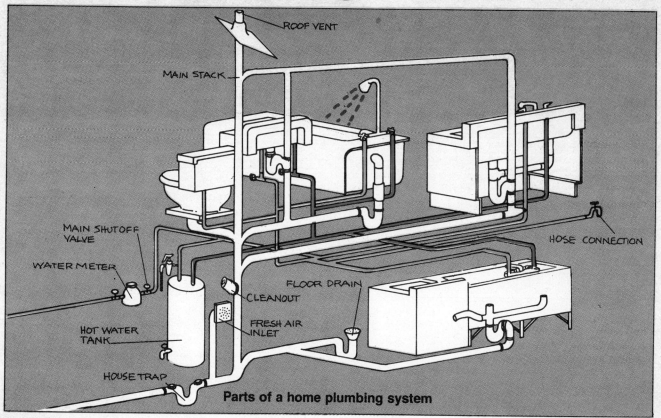

Parts of a home plumbing system

Getting started

If you know that water flows downhill and that a large-diameter pipe can carry more water than a small-diameter one, you're a long way toward becoming a do-it-yourself plumber, and saving some money.

In the old days, when leaded joints were the standard, house plumbing was limited to skilled journeymen. Today, with the introduction of plastics to plumbing, it is very much a do-it-yourself project. Even though your present plumbing system uses materials other than plastics, today's fittings provide a better, easier means of connection than the old leaded ones did. In short, you can do your own plumbing and carry the job as far as you wish. You can merely handle emergencies, such as a stopped-up drain or a pipe leak. Or you can go the whole route and install new piping, even putting in the plumbing for an added bathroom or a vacation cabin. It isn't our purpose in this book to turn you into a professional plumber. However, after reading it, and with some practice, you will be able to do quality household plumbing.

Every home-plumbing system features three parts: the water-supply system, the fixtures and appliances that use water, and the drain-waste-vent (or DWV) system. The water-supply system brings water into the house and distributes it to all the fixtures and appliances. These include hose attachments called bibbs, as well as fixtures like toilets, sinks, bathtubs, showers, and laundry tubs. Appliances consist of the water heater, water-treatment unit, dishwasher, clothes washer, furnace, humidifier, drinking fountain, garbage disposer, and boiler.

The drain-waste-vent system collects wastes from the fixtures and appliances and runs them out of the house into the sewer. It also is responsible for venting gases to the outside air, and is designed to provide access to the drain system, a necessity for clearing clogs.

For your purposes, the water-supply system begins at the point where water enters the house. This system consists of small pipes because the water in them is under pressure. The drain-waste-vent system ends 5 feet outside the house foundation, where the sewer begins. DWV operates on the principle of gravity flow. The pipes in this system are large since they must pass solids along with the liquid wastes. Fixtures and appliances are joined to one or both systems, depending on their use of water and their function.

To do your own plumbing, all you need is to learn the parts of the plumbing system, and accumulate a few simple tools. You probably already have some of them. If you plan to install any pipes, you'll need to know how they are joined, the different fittings that are available, and the limitations of the kind of pipe you will be using.

This book will give you all the know-how you need to do your own plumbing. Beyond that, the best teacher you can have is the job you do yourself. If you should run into problems, you can always ask questions of your local plumbing-supply dealer, hardware dealer, or home-center plumbing clerk. After reading this book, however, you may find that you know more than they do.

Another source of free advice— your plumbing inspector—will even make house calls. He will make the trip only to enforce local building codes, however, not to get on the end of a pipe wrench. All new plumbing you do will probably have to be inspected. Interior plumbing, whether on the water-supply or DWV system, comes under the jurisdiction of your city or county building-inspection department. Outside sewer building involves the local public-health department. These agencies will want to examine the finished work—before it is hidden from view—to see that approved methods have been used. Method and material requirements are both spelled out in your local plumbing code. Ask for a copy. A code is nothing to fear. It's for your protection and the protection of anyone who might buy your house later.

Many local codes adopt the provisions of the National Plumbing Code. Illustrations and descriptions in this book are based on the current code, which is probably available in your local library. If no local requirements exist, follow the provisions of the National Plumbing Code. Study it before you tackle a major job.

If your local code contains overly restrictive provisions that prevent you from using easy-does-it methods or materials—like flexible water-supply pipes—ask the plumbing inspector for what's called a *variance* or *experimental permit*. It's *your* house. If you want to use a simpler system than the local code allows, there's no good reason why you should not be granted a variance permit—as long as your system meets the standards of the National Plumbing Code. Although it is sometimes possible to avoid an inspection, much of what the inspector checks is what you would want to check anyway. Leak-testing is an example. If you work with your local inspector, you should find him more help than hindrance.

Some communities require you to apply for a permit to do even minor plumbing. In other communities, whether or not you need a permit often depends on what you call your project (see table below). If you term it *new work*, you'll need a permit. But if you can see your way clear to calling it maintenance, you don't need a permit. This can eliminate more than red tape. Your tax assessor looks at permits with an eye to increasing your property value, and thus your taxes. So, if you don't need a permit, don't get one. If in doubt, ask at your local buildings department.

What's the single most important thing to know about house plumbing? Where the main shutoff valve is located. That valve shuts off all water flow to the house. It's often in the utility room or crawl space, next to the water meter (if you have one). You may need to find it in a hurry someday. So all family members should be familiar with its location and be able to get to it on short notice.

Before a home water-supply system is approved, it usually needs a pressure test. For this, all system openings are capped or valved off and a pressure gauge is attached at one outlet. The water is turned on and then off at the main valve. If there are no leaks, pressure should hold steady overnight.

When a drain-waste-vent system is tested for leaks, all trap and fixture openings are sealed off. Then the system is filled with water at a roof vent-stack. The water level should hold in the stack overnight to prove out the system.

Home-buyer's checklist:

When you buy a house, your investigation should include the plumbing system. You want to know the quality of materials used and whether the system works as it should.

1. Throw a cigarette butt or crumpled piece of toilet paper into the toilet bowl and flush it down. If it doesn't disappear, the system may be sluggish and prone to stopping up.
2. Run a bathtubful of hot water and throw in some dimes. The water should be clean enough for you to tell whether they're showing heads or tails.
3. Thump on the bathtub. A loud, tinny sound indicates stamped metal—cheap. Likely the whole system is cheap. A more muffled, ringing sound points to cast iron—much better.
4. Turn on the hot- and cold-water faucets in all sinks and listen. The flow should be vigorous and there should be no piping sounds—rattling, creaking, groaning, and so on. If you hear them, it indicates restricted water flow.
5. Turn the faucets off rapidly and listen for "water hammer" (banging) in the piping. If you hear it, the system was probably built without protective air chambers. This points to poor design.
6. Look in the basement, crawl space, or attic to see what kinds of water-supply and DWV pipes were used. Compare them with the pipe charts on pages 44 and 49. From the standpoint of longevity, the best pipes are copper or brass for water-supply systems and copper or cast iron for DWV. Nothing wrong with the other types, though.
7. Note any evidence of water leaks on walls, floors, and ceilings.

Is a permit needed?		
Type of project	Typical jobs	Permit required?
Maintenance	Repairs, emergencies, replacing damaged or worn-out fixtures, replacing old piping.	No.
Remodeling	Replacing outmoded fixtures, modernizations, rearranging new or existing fixtures, adding fixtures.	Sometimes. Term it "maintenance" if possible.
Additional or new plumbing	Adding a bathroom, installing plumbing in a new garage, house, or vacation cabin.	Yes.

Water-supply system

Parts of the system

House-service entrance pipe. Brings water underground into the house.

Water meter. Measures the gallons of water used between readings so that the user may be billed for it.

Main shutoff valve. Cuts off water flow to the entire house.

Cold-water main. Routes cold water close to the fixtures and appliances that use it.

Hot-water main. Routes hot water close to fixtures and appliances that use it. Runs parallel to the cold-water main, about 6 inches away from it.

Branches. Fittings attached to the mains, that deliver hot or cold water to fixtures. The pipes are normally one size smaller than the main.

Fixture shutoff valves. Used in both hot and cold fixture-supply lines at the wall or floor behind the fixture. Permits selective shutoff of water to that fixture.

Air chamber. Used in both the hot- and cold-water lines behind fixtures and fast-shutoff appliances. Cushions water turnoff, preventing water hammer. A must for every system. Not needed on slow-shutoff fixtures such as toilets.

Temperature-and-pressure relief valve. Also called T&P valve. Used on the water-heater tank to release excessive steam pressure and heat. Prevents explosion. A must in all systems.

A house water-supply system is supposed to bring an adequate flow of clear, pure water to each fixture without much piping sound. And it should continue to do this job for a long time. The life of a water-supply system depends in great part on the pipe used. Some types resist corrosion and internal scaling better than others, and thus last longer.

Water for the supply system comes from a water utility—usually the city—or from a private water source such as a well. The quality of the water can vary from clear and delicious to murky and foul-tasting. In any case, the water *must* be potable—drinkable. If there is ever a doubt about potability, local health officials will test water free of charge. City water is tested for purity daily at the water-treatment plant. Private water sources should be tested at regular intervals, especially during wet times of the year, when contamination is most likely.

For full flow and pipe silence there must be pipes of adequate size throughout the water-supply system. A pipe that's too small, or one that has become scaled up on its inside surface, makes water speed up as it passes. This creates a disconcerting "water-running" sound. Using the right size pipe (see "Water-supply pipe sizes" table) ensures a quiet, efficient system.

In a piping system, lengths of straight pipe are joined by fittings. With fittings, you can have pipes to branch off, go around bends, and generally fit your plumbing plan. Since water flows through straight pipe much more easily than through angled fittings, the fewer fittings used in a system the better. Of course, a certain number of fittings are necessary just to make the installation.

The water in a house-supply system is under pressure—city water pressure runs about 50 psi (pounds of pressure per square inch; a standard measure). It is because of this pressure that the water-supply system can utilize much smaller pipes than a DWV system, and the water can flow up, down, or around as needed to reach fixtures. Gravity flow is important, however, should you need to drain the system, if, for example, the house is to be

Water-supply pipe sizes

Purpose	Pipe diameter
Fixtures	½ inch
Branches to fixtures	½ inch
Outdoor spigot	½ to ¾ inch
Cold, hot main	¾ inch
Service entrance	¾ or 1 inch

Some older houses use ⅜-inch supply tubing to toilets and sinks.

Main shutoff valve

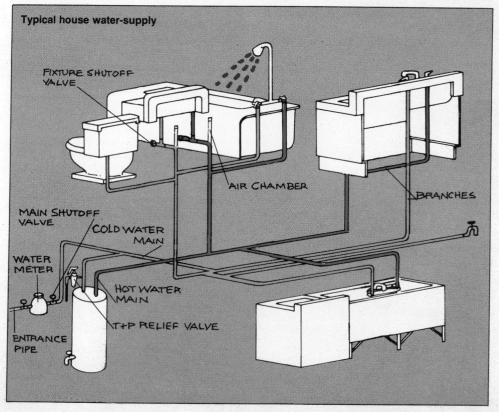
Typical house water-supply

unheated during below-freezing weather. Therefore, a well-designed water-supply system slopes slightly toward one or more low points. Drain or stop-and-drain valves located at those points are used to empty the system of water. Lacking these, other methods are available for cold-weather shutdown (see page 90).

When a water-supply system is built, direct pipe routings are chosen where possible. This is especially true of pipe runs that are hidden behind walls or in the attic. These are routed as directly as possible to reach their destination—across ceilings, through walls, and so on. With hot-water pipes, short direct runs are especially important. Shorter distances mean less heat loss—and saved energy. Where piping is exposed to view, however, as in a basement, most people want it to look neat and orderly. For this reason, visible pipes—even flexible-pipe installations—run with neat, square 90-degree turns.

The information above refers to the water system in general. With a detailed look, the system breaks down into three major types of pipes. The entry pipe, which supplies water to the house; the hot- and cold-water mains; and the branches that lead from the mains to the fixtures and appliances requiring water.

There are several types of branches. Parallel branches lead from the hot and cold mains to all fixtures that need both types of water. Hot water enters on the left side of these fixtures, cold on the right. Cold-only branches lead to the toilet and furnace humidifier. A dishwasher gets a hot-only branch. Sometimes a branch pipe serves more than one fixture. For example, single hot and cold branches may lead to a distant bathroom where they branch again to the various bath fixtures. This saves you the trouble of running mains to that room. In addition, smaller branch pipes have the advantage of delivering hot water more quickly, and with less heat loss, than main pipes can.

Some older water-supply systems have double piping between the water heater and bathroom fixtures. This maintains continuous circulation of hot water through the pipes and permits immediate hot-water flow from fixtures connected to them. Double piping is no longer installed because of its wasteful use of energy. The modern method is to insulate hot-water piping, and thus help it retain heat between water uses and in transit between the water heater and fixture. Page 42 shows the procedure for insulating a cold-water pipe to prevent condensation. The process is the same for a hot-water pipe, though the purpose is different.

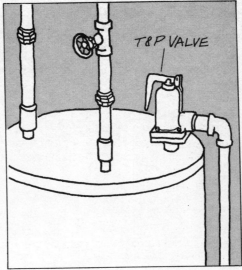

In addition to the water-delivering pipes, the water-supply system has a variety of other components. A temperature-and-pressure relief valve (T&P valve) on top of the water heater harmlessly bleeds away explosive pressures and temperatures.

To prevent annoying, potentially damaging water hammer, air chambers are placed in the wall behind each fast-shutoff fixture. These chambers usually consist of capped, 12-inch vertical lengths of pipe, of the same diameter as the water-supply pipe. As water flows toward an open valve, inertia prevents it from stopping immediately after the valve is closed. Instead, the water can "bounce around" viclently for a moment inside the pipe, causing the hammering sound. Severe hammering can build up enough force to split fittings, particularly in plastic-pipe systems. Air chambers harmlessly absorb the inertia and prevent both water hammer and damage. Fixtures, like toilets, that are not subject to quick shutoffs, don't need air chambers at all.

The last component of the water-supply system is the riser tube. One reaches from the shutoff valve (or adapter fitting) on each fixture to the faucet tailpiece. These tubes are often flexible, and they are used chiefly to simplify water-supply connections to fixtures.

All pipes of a water-supply system must be installed in such a manner as to be safe from freezing. When running pipes in exterior walls cannot be avoided, the pipes must be placed on the warm side of the wall's insulation. To protect pipes in unheated crawl spaces, either heat the crawl space slightly or wrap the pipes with electric heating cables. Pipe insulation is of little help in preventing freeze-up because, after a time, all heat escapes through even the thickest insulation. And insulation makes it more difficult to thaw out pipes that have already frozen.

How an air chamber works
When the faucet is turned off, water stops abruptly, compressing air in the chamber and cushioning the shutoff.

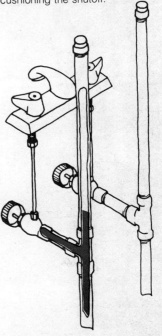

Fixture shutoff valve

BASICS 7

Drain-waste-vent system

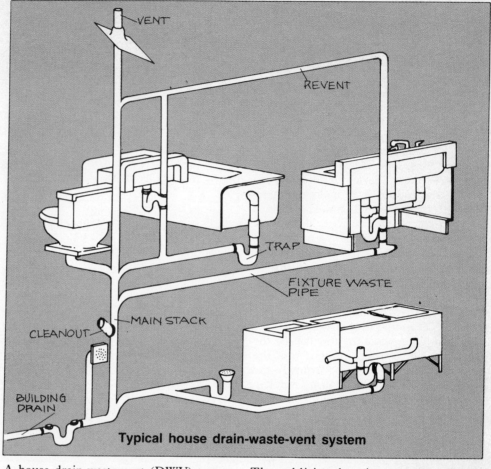

Typical house drain-waste-vent system

Parts of the system

Stack. A vertical pipe that collects wastes from fixtures. It vents up through the roof and is open at the upper end.

Vent. The upper portion of a stack that releases gases to the atmosphere.

Main stack. A stack that serves fixture drain and vent lines. Also called a soil stack.

Fixture waste pipe. Combination drain-vent pipe that connects a fixture or appliance with the rest of the DWV system. Sometimes called a branch drain.

Trap. Simple water-seal device between a fixture drain and the DWV system. Keeps sewer gases out of the house.

Revent. A vent-only run attached to a fixture to prevent trap siphonage. Rises from fixture and elbows into a stack above the highest fixture-waste connection.

Building drain. Drainpipe that collects all house wastes and leads them into the house sewer outside the foundation.

Cleanouts. Access openings in horizontal drainpipes, necessary for removing clogs.

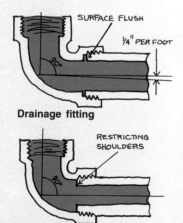

Drainage fitting

Water-supply fitting

A house drain-waste-vent (DWV) system carries fixture and appliance wastes away by gravity flow. Like a water-supply system, it should operate quietly, efficiently, and with proper flow. Some portions of the system, called vents, carry gases only. Others, called drains, carry wastes only. And some portions, called wet vents, carry both gases and wastes. Working together, these three components of the DWV system handle the remainder of the plumbing chores, picking up where the water-supply system leaves off.

Because DWV pipes and fittings are expensive, this system takes precedence in both the planning and installation stages. The rest of the plumbing system is planned around it. As many fixtures as possible are drained into a single main-drainage pipe, which is large enough to carry the entire flow. The idea is to plan your DWV system to use as little material as possible, yet not underpower it. Too big a system, besides being dollar wasteful, can be as troublesome as one too small.

DWV systems using lightweight pipes, such as copper and plastic, are likely to be noisier than those made of heavy cast iron. The additional noise usually doesn't amount to much. However, if there are sucking sounds as a fixture drains, there are problems with the DWV system's design or construction. Sluggish draining, or gurgling in one fixture trap as another fixture drains, likewise indicates troubles.

Used water coming from house fixtures contains solids. These tend to build up as greasy deposits on the insides of pipes and fittings. Build-ups are not serious in vertical drainpipes because the fast downflow of wastes scours them clean. But horizontal drain sections—that is, those sections that are only slightly sloped—can eventually get clogged. For this reason, every horizontal pipe must be accessible for cleaning. If the pipe is accessible through a fixture's drain, it may be cleaned from there. If not, an access point, called a cleanout, must be provided. Cleanouts are usually located at the higher ends of horizontal drainpipes, and they are usually covered with screw-on plugs. All cleanouts must be within easy reach, so when they're located in walls or ceilings, access doors should be provided. Sometimes access is through a toe-plug in the floor.

8 BASICS

Drain-waste-vent pipe sizes	
Purpose	Pipe diameter
Fixture waste pipes	1½ or 2 inches
Toilet waste	3 or 4 inches
Fixture vents	1½ inches
Toilet vent	2 to 4 inches
Vent increaser	4 to 6 inches
Building drain	3 or 4 inches
House sewer	4 inches
Some systems use 1¼-inch waste pipes for lavatories.	

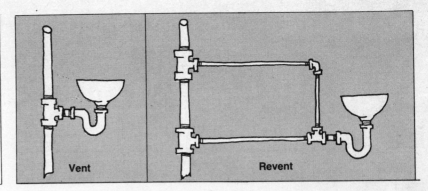

No cleanout should be located less than 18 inches from a wall or obstruction behind it. If this cannot be worked out, then the pipe is extended through the wall or obstruction and a cleanout fitting placed on the other side. Without the 18-inch clearance, getting a snake or auger into the fitting to clear a blockage may prove an impossible task.

Proper slope of the drainpipes is important. Too steep a slope lets liquids run ahead of solid wastes, leaving solids stranded in the pipe to form clogs. Too shallow a slope causes sluggish drainage and is also apt to clog. Properly sloped pipes drain quietly and efficiently and operate for long periods without clogs. A ¼-inch-per-foot slope is optimum.

Drain-waste-vent fittings are built differently from water-supply fittings, which need not work by gravity flow. DWV fittings are made with gently curving passages rather than sharp turns. The absence of shoulders in the fitting leaves no catch-points for the solids as they flow through. Scaled-down DWV fittings would work well in a water-supply system, but water-supply fittings, with their inner shoulders and obstructions, would soon short-circuit a DWV system. For this reason, even though water-supply fittings are less expensive, they should not be used for the DWV system. Of course, the pipe and fitting sizes are different for the two systems, and mixing them would be tough, in any case.

Drainage fittings may be used throughout a DWV system. But in portions where no waste water is present—the vent portions—it is better to use vent-type fittings. These fittings do the same job as drainage fitting, but cost less.

Every fixture must have a trap (see drawing opposite). The one in the space under your kitchen or bathroom sink is probably the most familiar. These traps are usually the same size as the waste pipe—often 1½ inches—but can be scaled down to 1¼ inches. Avoid using a trap with a larger diameter than the waste pipe, as it may cause clogs. The best traps are those with built-in cleanout plugs.

Though you can't see the traps on most other fixtures, they all have them. Bathtub, shower, and automatic-washer traps are underneath the fixture, below floor level. Dishwashers and garbage disposers share the kitchen-sink trap, while every toilet has a built-in trap.

If it weren't for these traps, another part of the DWV system, the vents, would not be needed. Vents provide an outlet for gases, and keep air pressure from building up. This pressure could siphon fixture traps dry, making them useless. Thus every fixture must be vented. Venting extends the DWV system upward through the house and opens it to the atmosphere, 12 inches above the roof.

Some fixtures perform both draining and venting tasks through their waste pipes. These pipes are known as wet vents. If wet vents have to travel too far (between the fixture's trap and the vent stack), extremely rapid waste flow, and siphonage, can occur. To avoid this, revents are used. These are simply vents that extend upward from a fixture and join the vent stack higher up. Fixtures with small waste pipes are especially subject to wet-vent siphonage and should be placed close to the vent stacks that serve them. Otherwise, revents must be installed. Reventing a toilet, with its large drain pipe, should be avoided. If the toilet is within 24 inches of a main vent, no reventing is needed. If this placement is impossible, a specialized treatment for a toilet revent is necessary. With a Y-fitting, the toilet waste pipe is extended "upstream" from the toilet. There it is connected with a secondary stack that is the same size as the waste pipe, usually 3 inches. This is hardly reventing at all, since an entire new stack must be extended through the roofline. So, try to design toilet locations near main vent-stacks. In fact, reventing in general, because of the additional pipe, fittings, and time it requires, should be avoided if possible. Do so by placing fixtures close to vent-stacks. A few very restrictive local codes, however, still prohibit wet vents and make it necessary to revent all fixtures.

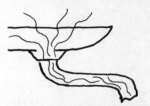

Unsealed drain would let gases into room.

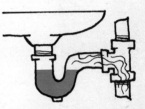

Gases sealed off by drain-trap water.

Unvented trap would be sucked dry by siphon action.

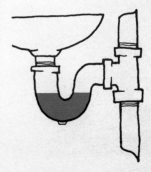

Vented trap is not siphoned so water seal stays in place.

Fixture and appliance hookup

If you add a bathroom or a new appliance, you'll be working with fixture plumbing. The following paragraphs will give you a general introduction to installing fixtures and appliances. The various jobs have much in common. The table of contents for this volume shows the page numbers where instructions for each type of installation will be found.

Bathtubs and showers drain below the floor, either through a P-trap or a drum trap. The P-trap may be cleaned through the tub's drain. A drum trap has a removable cover. This must be accessible below the floor or through a floor plate.

For a shower, the piping is installed first. Then the shower floor pan is laid over it and the drain caulked. For bathtub installations, you need access from below the tub once it is in place. Where bathtub access must be from above, as in a concrete-floor house, it's provided through a wall-access panel at the head end of the tub. Trap connections are made in a 12-inch-wide, 14-inch-long galley in the floor below the tub's drain. A toilet connects to its closet flange without underneath access (see page 35).

Water-supply connections to tubs and showers cannot use the flexible riser tubes shown at left because of the greater (5-gallon-a-minute) flow required. (Most fixtures need only about 3 gpm.)

Most tub and shower mixing controls are designed to take ½-inch threaded pipes or adapters. Some contain ½-inch sweat-solder fittings for a direct copper-tube hookup. A giant-sized escutcheon, or shield, covers the hookup hardware.

In framing for fixtures, be sure to provide headers—horizontal 1-by-4-inch boards nailed into the framing. They are necessary supports for lavatory sinks and also hold the lavatory water-supply outlets. They're needed for tubs, too, where they secure the faucet and control, the shower arm, and the tub-rim supports. See the drawing on the next page.

Lavatories and sinks drain through separate P- or S-traps. Use P-traps for wall waste pipes, S-traps for floor waste. A wall-waste setup is preferable.

Water-supply hookups to sinks and lavatories are made with fixture shutoff valves and riser tubes. Use angle-stops from walls, straight-stops from floors.

A water heater needs an energy connection and, if it's a fuel-fired heater, a vent. Plumbing connections consist of a valved cold-water inlet and a hot-water outlet. A heater also requires a T&P valve with a ¾-inch relief pipe, which leads to a floor drain or other convenient emergency-disposal spot, as shown at right.

A water softener is connected in series with the cold-water main. Easiest to use are flexible connectors designed for that purpose.

Garbage disposers are mounted in kitchen-sink drains between the sink and the fixture trap. Besides the plumbing hookup, a disposer requires an electrical connection. Be sure to follow the manufacturer's instructions for this.

A dishwasher—if not a portable—gets a hot-water-supply pipe all its own. The simplest connection is made with flexible tubing and a flare fitting at the dishwasher inlet valve. Be sure to provide a shutoff valve in the line. Dishwasher wastes are pumped out through a flexible rubber tube into a side tapping on the sink drain.

An automatic washer often utilizes an existing laundry tub with its hose bibbs and drain features. It can also be installed with separate hot and cold hose bibbs of its own. Drainage then is into a 1½-inch standpipe that reaches 36 inches above the floor. This standpipe should have a below-floor trap.

Using riser tubes

Flexible metal riser tubes can be cut to size and bent by hand to fit between fixture and its water-supply shutoff valve, eliminating the need to line up the two pipes perfectly. Tube with bullet-nosed end inserts into sink or lavatory faucet tailpiece; tube with flat end connects to toilet tank inlet pipe.

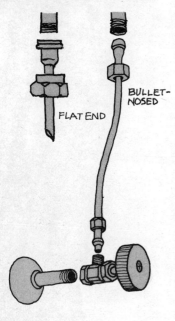

Using trap adapters

Slipjoint trap adapter connects fixture trap to waste pipe quickly. Adapter slips over trap arm which can be slid in or out as needed to reach fixture drain. Then trap slip-nut, with O-ring washer inside, screws onto adapter's outer threads to form gas-tight, water-tight joint.

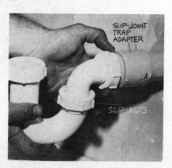

New wrinkles

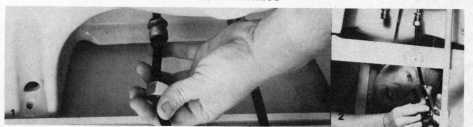

A newly marketed, flexible polybutylene riser tube bends easily to reach between the wall or floor fixture-shutoff-valve and the faucet tailpiece. It withstands hot and cold water; bends without kinking. The new riser fits standard ⅜-inch fixture shutoff valves. To install it, (1) cut the riser to size with a knife and slip the compression nut and brass ferrule on the riser, (2) install the riser in the shutoff valve and tighten the compression nut.

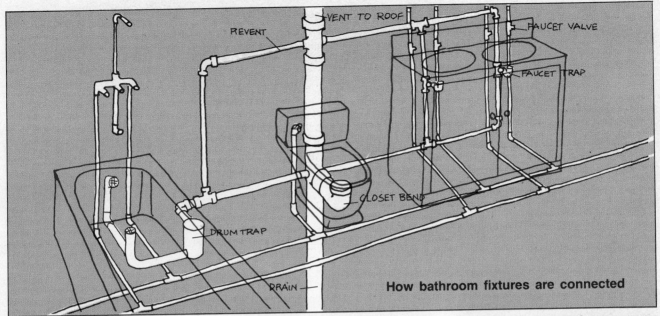
How bathroom fixtures are connected

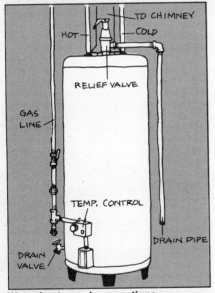
Water heater and connections

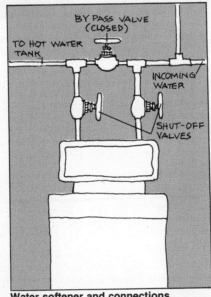
Water softener and connections

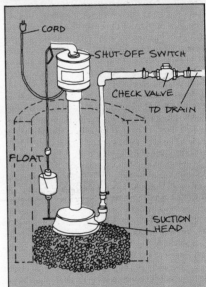
Sump-pump connections

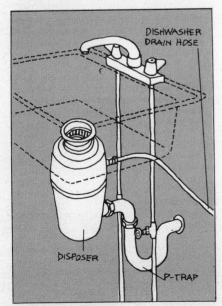
Garbage disposer

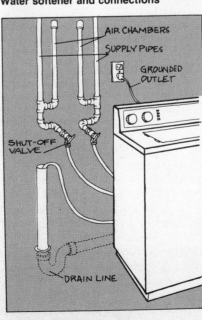
Automatic-washer hookup

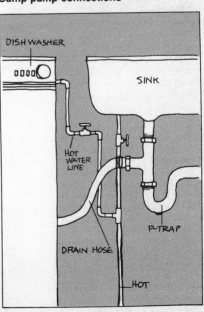
Dishwasher connections

BASICS 11

Plumbing tools

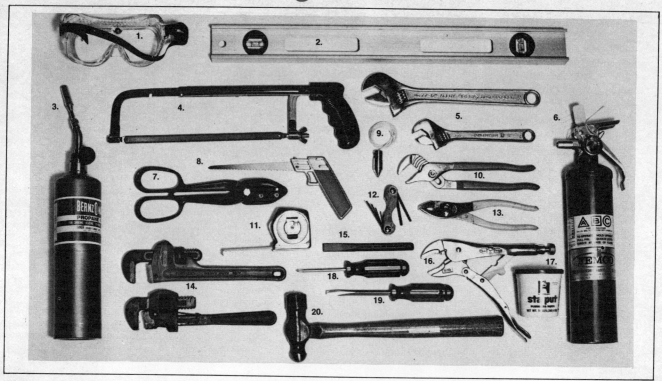

(**1.**) Goggles or safety glasses $2–$6. (**2.**) Level $6–$10. (**3.**) Propane torch $8–$18. (**4.**) Hacksaw $4–$7. (**5.**) 6- and 12-inch open-end adjustable wrenches $5–$10. (**6.**) Fire extinguisher (ABC rated, with gauge) $13–$30. (**7.**) Metal or aircraft snips $4–$7. (**8.**) Keyhole saw $2–$4. (**9.**) Plumb line $1–$2. (**10.**) 10-inch channel-pliers $5–$9. (**11.**) Retractable measuring tape $5–$12. (**12.**) Allen-wrench set (for setscrews) $3–$5. (**13.**) 8-inch slip-joint pliers $5–$7. (**14.**) Two pipe wrenches $7–$11 each. (**15.**) ¾-inch cold chisel $3–$4. (**16.**) 10-inch locking plier/wrench $4.50–$7. (**17.**) Plumber's putty $1–$2. (**18.**) Phillips screwdriver $2.50–$4. (**19.**) 10-inch screwdriver $3–$5. (**20.**) Ball-peen hammer $7–$12.

Basic tools used for plumbing (see above) may include many you already own. They belong in every home workshop. If the list includes any tools you don't have, consider acquiring them. It's a lifetime investment.

In order to do any actual work with plumbing pipes and fittings, you also need most of the tools in the general-plumbing category. Here you can be selective. Choose tools according to the type of pipe you'll be installing. Only for systems using No-Hub drain pipes, for instance, would you need to buy the No-Hub torque wrench.

Maintenance tools are often acquired only when a problem arises. Basic to plumbing maintenance, and the first tool to reach for when a clog occurs, is the plumber's force cup (see page 38). When the clog is tougher, maintenance tools with a bit more clearing power can be bought. But these tough-clog clearers and other tools are often too costly to warrant ownership for just occasional use. Most can be rented on an hourly or daily basis.

How to use basic tools

A **claw hammer** and a **ball-peen hammer** are not interchangeable. Use a claw hammer only for driving and pulling nails. If you pound on a cold chisel with a claw hammer, you risk chipping its face and ruining it. A ball-peen machinist's hammer has a harder face and is designed for this use. Grip a hammer close to the end of the handle. Don't "choke" it. This improves your leverage; it also helps to keep the face of the hammer flat to the work.

There's a right and a wrong way with other tools, too. A **screwdriver** should fit snugly into the screw slot, with little or no play. Keep the shaft of the tool vertical to the head of the screw. The 10-inch screwdriver is for driving or removing standard slotted screws; the other is for Phillips-head screws. (No. 2 is the most common Phillips screwdriver size. It will also handle the small No. 1 Phillips-head screws found on some faucet handles.)

Pliers are among the most often misused tools. They were never intended to take the place of a wrench. The hardened jaws of a pair of pliers, when used on a soft brass or plastic nut, will chew it up. Pliers can also round off the corners of harder nuts, sometimes making them unremovable. Pliers are great when you need more leverage than your own hands can supply. But a good rule to follow is to never use pliers if another tool can do the job as well. Most standard pliers are of the

Get sharp
It didn't take me long to get the point. Now I believe in keeping tools sharp. You not only work faster, you work safer. Properly sharpened tools cut well without being forced. It's when you have to force a tool to do its job that you get bruised knuckles, scraped hands, and worse. So keep a proper edge on chisels, screwdrivers, knives, and saws. And if the gripper jaws on a pipe wrench get rounded off after long use, get a new wrench. It will pay off in saved knuckles.

Practical Pete

slip-joint type. These can be set at two widths for gripping narrow or wide objects.

Similar to the slip-joint pliers, but with a greater range of width settings, are **channel-joint pliers**. A good pair may be substituted for a pipe wrench if the turning isn't too hard.

Completing the common plier types are **locking pliers**. These feature adjustable jaws. Properly adjusted, and with the handles squeezed together, locking pliers form a portable vise. As such, they can be used for holding pipes when you need to saw them, or for temporarily securing a part before it is permanently attached. Use locking pliers with care, however, because they have enough leverage to deform copper tubing and fittings. The best locking pliers have an easy-release lever that lets you remove them without pulling on the handles, which would otherwise be hard to open.

Longer and longer lengths have become the trend in **measuring tapes**. The 25-footers are now commonly used, meaning that very long measurements can be taken directly. Adding up shorter lengths isn't necessary. Handiest of all the measuring tapes are those with a spring-loaded, self-return mechanism. For plumbing jobs where long measurements aren't often needed, a less expensive yardstick will do.

The **carpenter's handsaw** is used for notching framing members to make way for pipes. It can also be used—in conjunction with a miter box—to cut plastic pipe. (The miter box is necessary for getting a square edge.)

If you have an old saw that's ready to be thrown out, save it for cutting pitch-fiber sewer-septic pipes. These will leave black stains on any saw used to cut them. If you don't have an old handsaw, remove the stains from your good saw with solvent. Then oil it to protect from rust.

Metal pipe should be cut with a **hacksaw**. Slice through thin-walled copper tubing with a fine (32-teeth-per-inch) blade. For other cutting, use a regular 24-tooth blade. A hacksaw works just as well as a carpenter's handsaw on plastic pipe.

Where holes are needed in tight quarters, as for a pipe run, a **keyhole saw** does the job. Its narrow, pointed blade allows you to start a cut in a drilled hole. The handiest type has removable, reversible blades in different sizes. Put in backwards, the blade can saw right up to a flat surface, letting you cut through a pipe that runs along a wall.

Open-end **adjustable wrenches** are used for turning nuts up to 1¼ inches in width. Use the 6-incher for small nuts, like those found on faucets, and the 12-incher for larger ones, as on valve bonnets. Always position adjustable wrenches so that you are working the handle toward—not away from—the movable jaw. This keeps the stress on the strong cast housing, rather than on the weaker lower jaw. An old-fashioned **monkey wrench**, which has the opening on the side instead of the end, can be substituted for either or both of the open-end wrenches.

Allen wrenches are used for setscrews which are sometimes found on faucets and their escutcheons, or shields. These wrenches have closed, hexagonal heads that can be inserted into the recess of a setscrew to turn it. Allen wrenches usually come in sets. One handy version includes most of the smaller sizes in a single tool, with the wrenches hinged like the blades of a pocketknife.

Most **propane torches** come in kit form, complete with various tips that give small or large flames. You'll want a large tip for sweat-soldering copper pipes. When you turn off the torch between uses, be sure that the flame goes out. Unscrew the torch from the canister between jobs, and follow the storage cautions on the container. You'll find that a flint striker is much handier than matches for torch lighting.

The **cold chisel** made completely of metal, has many uses in plumbing. It's used for severing cast-iron pipes (see page 49), and chewing through stubborn nuts that won't come off. Any time you need real power for cutting through metal, a cold chisel can provide it. Pound on this chisel only with a ball-peen hammer, and always wear goggles when you use it. A chisel whose head is badly "mushroomed" from long use should be ground flat, with a slightly beveled edge. Otherwise, flakes of metal may dislodge. The tip may occasionally have to be restored too. When you do this, be careful not to change its original angle. For both jobs (known as redressing) use a grinding wheel and wear eye protection.

You'll need a **level** for most plumbing jobs. The 2-footer recommended in the tools list is best. Look for one with three bubble tubes. These are opposed to one another, enabling you to check for level horizontally and for plumb vertically.

Whenever you work with hand tools, minimize all possibilities for injury. A pair of **work gloves** can save splinters and blisters, and **goggles** or safety glasses are a must whenever loose particles or hazardous liquids that may enter your eyes are present. A **fire extinguisher**—the dry-chemical type—should be a part of every tool kit. One is especially vital when you are working with a propane torch.

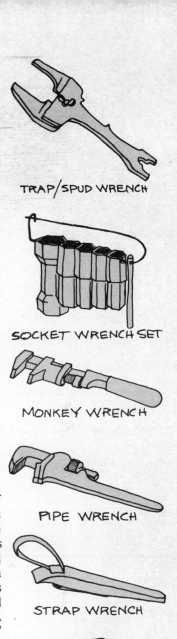

TRAP/SPUD WRENCH

SOCKET WRENCH SET

MONKEY WRENCH

PIPE WRENCH

STRAP WRENCH

TUBE CUTTER

BASIN WRENCH

Tips for general plumbing tools

Having **pipe wrenches** in two sizes, as noted in the tool list on page 12, is helpful in many plumbing jobs. Either may be used singly for turning threaded pipes and fittings when the opposite end is well anchored. They can also be used as a pair when one end is not anchored (see illustration below). A small wrench is best unless you need greater leverage than it provides. Where this added leverage is needed, go to a larger wrench. Never put anything over the handle of a tool to increase its leverage. Pipe wrenches will bite in only one direction and only when the handle is pulled toward the movable jaw. Pushing the other way removes the wrench, or lets it alternately bite and release for a kind of ratchet action.

On exposed pipes, where you want to avoid making teeth marks, use a **strap wrench.** This has a heavy nylon strap instead of jaws. To use one, start with the strap coming out the convex side of the wrench, as shown in the illustration below. Wrap the strap around the pipe. Then feed it back into the slot in the wrench. If properly installed, the strap will tighten as you pull on the handle.

Use flame with care

What's the worst possible time for a house fire to start? When the water supply is turned off! Remember that fact whenever you are soldering. A torch flame can easily start a house fire. If it gets sawdust smoldering, a fire can break out later. Carry a small dry-chemical fire extinguisher with you wherever you solder. And when you solder a fitting that's close to wood, always put a sheet of asbestos behind the fitting to stop the flame. Soaking the wood behind the fitting with water helps, too. Be careful where you direct the torch's flame while you're concentrating on feeding solder into a fitting. And, of course, don't use a flame around combustible liquids or gases.

Practical Pete

Tighten fittings with two pipe wrenches, one to hold and one to turn. Apply force toward the jaw openings.

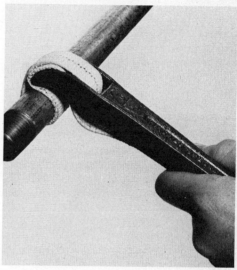

Use a strap wrench to avoid leaving teeth marks on soft pipes such as brass.

Water-supply pipes, whether steel, copper, or plastic, can be cut with **pipe or tubing cutters.** These have sharp wheels that neatly slice through pipes. To use a cutter, open it wide and place it over the pipe. Position the cutting wheel over your mark. Tighten the wheel securely, then revolve the cutter around the pipe until it cuts through the wall. After each full revolution, you'll have to retighten the clamp. Many tubing cutters contain a built-in **pipe reamer,** making separate purchase of that tool unnecessary.

Flaring tools are used in conjunction with flare nuts to make flared joints in soft-tempered copper tubing. These tools are of two types: drive-in and clamp-in. A drive-in flarer is simply a cone-shaped device that is driven into the end of the pipe until the desired flare is created. Clamp-in flarers have two parts, a vise bar that holds the pipe and a clamp with cone-shaped die that makes the flare. The farther the pipe protrudes from the vise, the bigger the flare. Whichever type you use, always put the flare nut on the pipe before shaping.

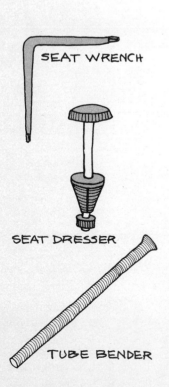

Cutting wheels in pipe and tube cutters slice into pipe or tubing as cutter is rotated. After each circular cut is made, cutter is tightened slightly.

To flare a tube end, slip flare nut on tubing, insert tube end in opening the same size in die, tighten wing nuts, insert flaring tool tip in center of tube, and tighten by turning clockwise.

For working in tight quarters underneath and behind fixtures, you'll need a **basin wrench**. It has a long arm that can reach behind a lavatory to loosen or tighten fixture nuts on the water-supply tubes. An **adjustable trap** or **spud wrench** is for use on large nuts such as those on sink strainers, traps, and toilets. Another fixture tool, used for installing or removing faucets, is the **faucet spanner**. One usually comes with a new-faucet installation kit.

Most well-stocked rental centers carry **drain augers**, **rotary root cutters**, and **pipe-threading equipment**. These relatively high-cost items should be rented unless steady use is anticipated.

Pipe threaders are used only for metal water-supply pipe, not plastic. Most come with an assortment of dies, as well as the pipe stock (handle) and vise. The important thing to remember is to use plenty of thread-cutting oil to keep from chipping the newly-formed threads and to prolong die life (see page 46).

Usage of specialized tools not described here will be covered in later sections.

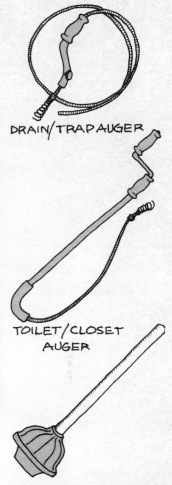

DRAIN/TRAP AUGER

TOILET/CLOSET AUGER

PLUMBERS FORCE CUP

A basin wrench gets into snug quarters underneath sinks and lavatories.

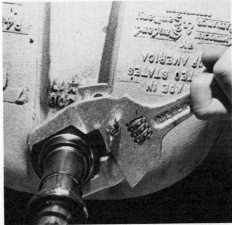

Adjustable spud or trap wrenches fit large slip nuts such as those on toilet tank connections.

A faucet spanner, provided with new faucets or repair kits, is used to tighten or loosen faucet adjusting ring to prevent stem leaks. Get one that matches your model faucet.

Tools for plastic pipe

Working with easy-does-it plastic pipe takes very few tools beyond those shown in the illustration on page 12. The only additional items you'll need are: a 1½-inch paintbrush; 9/16- and ¾-inch hexagonal Allen wrenches (for tightening transition unions); penknife and sandpaper (for smoothing pipes). One of the tools shown—the tubing cutter—is optional. If you use one, get one with a plastic-cutting wheel. But, you can substitute a handsaw. It works as well as the cutter.

Not much of a tool kit. But it's about all you'll ever need for plastic pipe.

A chip off the old block
A few years ago, I used to laugh at the idea of wearing goggles. "Not me," I'd say. Then, one time when I was sharpening a chisel on a grinder, a chip of metal nearly flew into my eye. That taught me a lesson. Now I wear safety glasses or goggles whenever there's any chance for eye injury. Be smart. Protect your eyes whenever you are grinding, drilling, chipping, soldering, using a chisel—for any job where a particle or liquid could get into your eye. Believe me, it can happen.

Practical Pete

When soldering or thawing pipes always place an asbestos sheet between the flame and nearby beams or wall surfaces. To solder a joint, heat the fitting and pipe until tip of solder, touched to opposite side of joint, melts and flows into the joint. Don't heat solder.

2. FAUCETS

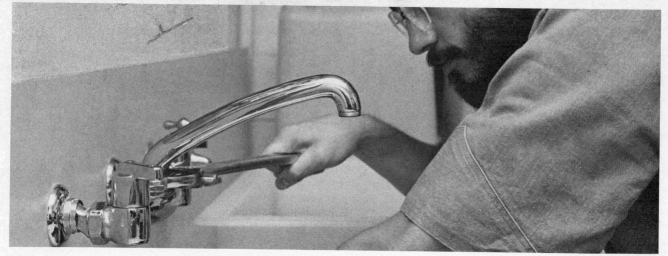

Faucets are valves attached to the ends of pipes. Fixing them can vary from simple washer replacements to installing new spindle assemblies. Replacement parts are available separately or in kits.

Buy parts that match your type of faucet. There are several types. Simplest of all, and the original one, is the washer-type faucet.

Washer-type faucets

Despite the increasing popularity of washerless faucets, many washer faucets are still around. And although they require more frequent repairs, they are usually the simplest to fix. These faucets work by means of a rubber washer attached to the faucet spindle. When the handle is turned to the *off* position, the spindle is lowered, and the washer is clamped tightly onto a metal seat. This closes off water flow through the faucet body. The clamping is done by fast-acting threads on the spindle.

This is the type found in most old plumbing systems, and is the least expensive faucet to buy. In more modern systems, you are likely to find washerless faucets—units that use diaphragms, discs, valves, balls, or cartridges in place of washers. More costly but less troublesome than washer-type faucets, these are becoming standard.

It's a beautifully simple system, but rapid wear results from the closing action because the washer is twisted as it is pressed onto the seat. In time, especially on the hot-water side where the washer is softened by heat, you must turn harder to stop all flow. This causes still more wear, and ultimately, when the faucet no longer can be shut off completely, you must repair it. As soon as you have to force the handle to stop the flow, get set to make the repair.

(continued on page 18)

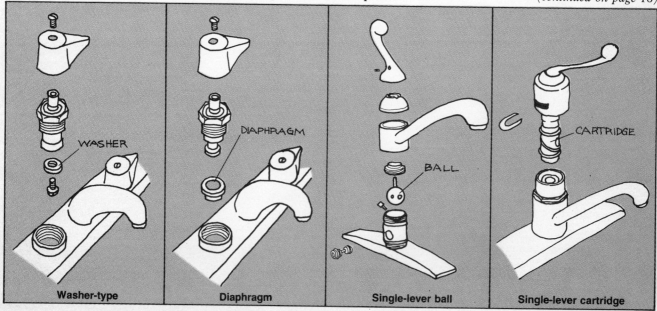

| Washer-type | Diaphragm | Single-lever ball | Single-lever cartridge |

Replacing a washer

1. To replace a leaking faucet washer, first turn off the water at the fixture shutoff. Remove the handle, escutcheon, if any, and other parts to expose the packing nut.

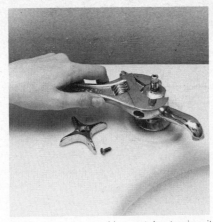

2. Unscrew the packing nut by turning it counterclockwise with a wrench. Some nuts seal around the faucet spindle with packing. More modern ones use rubber O-ring seals.

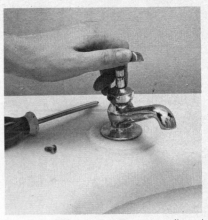

3. By reinstalling the handle temporarily and turning it in the *on* direction, you can thread the spindle out of the faucet body and lift it free. Examine it for damage to O-rings, packing, and washer. Replace damaged parts, including the washer.

4. If the old washer is abraded, flattened, or has become hard, replace it with a new one of the same size. To take it out, remove the brass central screw and pry it loose. Your dealer can probably match it up with a new washer and sell you a new brass screw, if replacement is necessary.

5. Install the new washer on the spindle, flat face in. In a pinch, you can install the old washer backward as a temporary repair. Inspect the faucet seat for nicks, scratches, abrasion. Some worn seats can be threaded out for replacement. Others must be dressed smooth with a seat-dressing tool (see page 18).

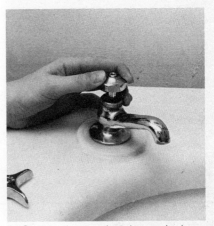

6. Once you have replaced or repaired worn parts, reassemble the faucet in the reverse order in which you took it apart.

If faucet packing needs to be replaced, use Teflon packing. With the packing nut removed, wrap the new material tightly around the faucet stem, winding clockwise, until it is thick enough to be compressed. Then replace the nut. This draws the packing against the stem.

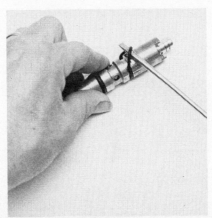

To replace O-ring, pinch it to raise part of it, insert knife or screwdriver blade under raised part and roll the ring off. Roll on new ring. New ring must match old one in size, so take old ring or spindle along when buying.

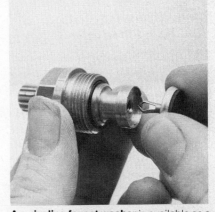

A swiveling faucet washer is available as a direct replacement for the original screw-on one. Its expanded prongs hold it in the empty screw opening. As the faucet is turned off and the swiveling washer contacts the faucet seat, it turns freely instead of wearing on the seat. File off the spindle's shoulder before installing the washer.

What it takes

Approximate time: Rarely will it take more than an hour to fix a faucet unless you have to run out for parts.

Tools and materials: You'll need a standard screwdriver, a Phillips-head screwdriver, and an adjustable wrench. If the faucet has an escutcheon held by a setscrew, you'll need a setscrew wrench that fits it. Depending on the faucet, you may also need a kitchen knife, and a seat-dressing tool (or a valve-seat removal tool). Repair materials vary depending on the kind of faucet and its make. For a washer-type faucet, the minimum will be a new washer. The most would be a new spindle and a seat-replacement kit or new seat. For washerless faucets, you'll need a basic repair kit, a more complete rebuilding kit, or a new spindle or cartridge at the most. It's best to get the faucet apart before buying any parts.

Planning hints: Do the job when your plumbing-parts dealer is open, so you won't have to wait for parts. If the faucet is not served through a fixture shutoff valve and you'll have all the house water turned off, plan the job for a time when water will be least needed by other family members.

That beats waiting until the faucet seat is damaged. It's a good idea to fix *both* sides of a washer-type faucet even though only one side requires it.

When you replace a washer, you may see that the brass screw that holds it in place is worn or corroded and in need of replacement. Renew it with another brass screw, never with one of a different material. The right size screws, along with other faucet-repair parts, are available at your plumbing-supply store. The surest way to obtain the correct replacements is to get the manufacturer's name from the faucet body—(new parts are designated by faucet make and model)—and then take the old part in with you.

Washer-type faucets come in two forms: the packed-stem type and the more modern O-ring-sealed type. The names refer to the method used to keep the faucet from dribbling around the handle stem when the faucet is turned on. Packing is graphite-impregnated cord, a material that looks like fine black spaghetti. O-rings do a more effective job of sealing than packing does. They also create less friction on the valve stem and are more easily replaced. An O-ring-sealed stem can actually be loose in the faucet body, yet remain leak-free. You can get new O-rings or a complete spindle assembly with the washer installed.

A good many washer-type faucets have removable seats. It's often easier to remove and install a new seat than it is to repair a worn one. Once you get the faucet's spindle out, it's easy to tell whether or not the seat is removable. If it is, it will have a square or hexagonal hole through its center. If the seat is an integral part of the faucet body, it will have a round hole. Insert a seat-removal tool—sometimes a large setscrew wrench will do—into the hole, and turn it counterclockwise to thread the seat out of the faucet. Turn the new seat clockwise to install it. If you lack the proper seat-removal tool, an alternative is to find a screwdriver blade that will fit tightly across two corners of the hole. Tap it in, then grasp the screwdriver handle with both hands, and turn it hard to loosen the seat. If that won't work, break a tool-use rule and try turning the screwdriver with pliers. If you install a new seat with the somewhat brutal screwdriver method, be sure not to jam up its threads or damage its seating surface in the process.

To install a new seat in a faucet that doesn't have removable seats, get a seat-replacement kit. This method, cheaper than replacing the entire faucet, creates a space in the old seat where a new stainless-steel seat can be glued. It works on laundry, tub, and shower faucets as well as deck-style (counter- or sink-top mounted) faucets. In fact, the repair methods and parts described in this section apply to all household faucets.

With all the talk about faucet seats, something more on faucet washers is in order. Washers come in many sizes, from ¼ inch up. They're made of fiber or medium-hard neoprene rubber. Those for faucets should state on the package that they're *faucet* washers, meaning that they are made to withstand hot water as well as cold. Valve washers, though they look much the same, may not be able to take heat. One side of the faucet washer is flat; the other side is flat, tapered, or rounded. Flat washers are used with crowned and ridged seats, usually replaceable ones. Always install the washer with a flat side against the spindle. In time, all washers become hardened and compressed. The seat forms a groove in them. That's when

Fixing valve seats

1. With the spindle out, most valve seats can be removed with a seat wrench or screwdriver. Smooth slightly worn seats with a seat dresser; replace badly worn seats.

2. Fit seat dresser into faucet body and turn to dress the old valve seat smooth.

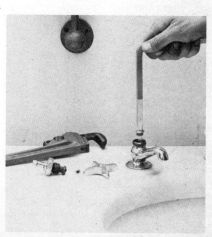

3. Get a matching replacement seat, coat its threads with pipe joint compound, fix it to seat wrench tip, and screw it into the faucet body. Then reassemble the faucet.

Disc faucet repair

1. To fix a leaking disc-type faucet, remove the spindle assembly from the faucet and replace it. Sometimes it takes quite a pull to get the cartridge out (if it isn't threaded in). For added leverage, reinstall the handle.

2. Some disc-type faucets with ceramic cartridges are held together with screws. Remove cap and handle, and the two screws. You can repair this kind by installing new O-rings around the inlet and outlet and spout holes in the bottom. Remove the screws and lift out the parts to get at the O-rings.

Lever faucet repair

1. Single-lever faucets are easily repaired with kits. To repair a dripping faucet, all you need is a basic seat-and-spring kit. Also available are complete parts kits that can be used to stop dripping and leaking around the control handle. Directions are included with the kits.

2. You can stop most handle leaks simply by screwing down the adjusting ring with a spanner wrench that's part of the repair kit. Tighten until you feel a slight drag as the stem is moved back and forth.

3. To repair a dripping faucet, replace the seats. First take faucet apart and discard the old seats and springs. Install new ones, springs down. Reassemble. Be sure to reset the adusting ring tension properly.

4. To stop a leak coming from beneath the bonnet cap, replace its seal (from the kit). Cut the old O-ring seal to remove it, stretch a new one, and snap it into the groove. Reassemble the faucet. If your faucet is built differently, write the manufacturer for an instruction sheet.

they're likely to leak. If it happens to a new washer in a short time, a roughened seat is probably scoring the washer.

A faucet that vibrates or chatters as water flows may have a loose washer or worn threads on the spindle. Tightening the washer or replacing a worn spindle can stop the noise in most cases.

If a faucet handle is lost or broken, it can be replaced. If you cannot find a duplicate of the original handle, try replacing both handles with a matched pair that has the right kind of splines or strips to fit your faucet's spindle. When you reinstall a handle, first put it on the shaft loosely, without using the hold-down screw. With the water supply on, turn the faucet on and off a few times. This will establish the *off* position, enabling you to remove the handle and replace it on its splined shaft in neat alignment. Install the hold-down screw and cap and you're done. Should the *off* position change as the washer wears, simply take the handle off again and replace it in proper alignment. You may find that alignment on the hot side changes faster than on the cold side.

Though they are simple and their workings are easy to understand, washer-type faucets are more prone to problems than washerless because they have more parts that can wear out.

Washerless faucets

Washerless faucets are of many types. The main visible difference between the types is that some have a single handle and others have dual handles. All washer-type faucets have a pair of handles, one controlling the hot-water flow and the other the cold. But some washerless faucets also have dual handles and, externally, they closely resemble the washer type. The handles of a washer-type faucet move up and down, if only slightly, when they're opened and closed. This happens as the spindle, with attached washer, threads in and out. With a washerless-faucet handle, there is no visible up-and-down movement. This distinction is important because repair parts for the two faucet types are completely different from one another.

Discovering that a dual-handled faucet is the washerless type is good news, for the most part. Washerless faucets rarely need replacement parts. The bad news is that

Hidden handle tricks

Faucet manufacturers are experts at hiding just how faucet handles are held to their shafts. For appearances, the screws are often secreted beneath trim caps on the faucet handle. Common is the snap cap. To remove this type, work a knife blade under the cap and pry it up gently. Don't use one of your best knives. It's likely to get bent. Other caps are threaded in. You can try unscrewing this type simply by turning it counter-clockwise with your fingers. If that won't budge it, tape around the cap and turn it with pliers. Be careful not to break, bend, or mar the cap. Once you get it off, the rest is easy. Simply remove the screw and lift the handle off its shaft. Whatever you do on a faucet, don't force anything.

Parts of a kitchen faucet with sprayer

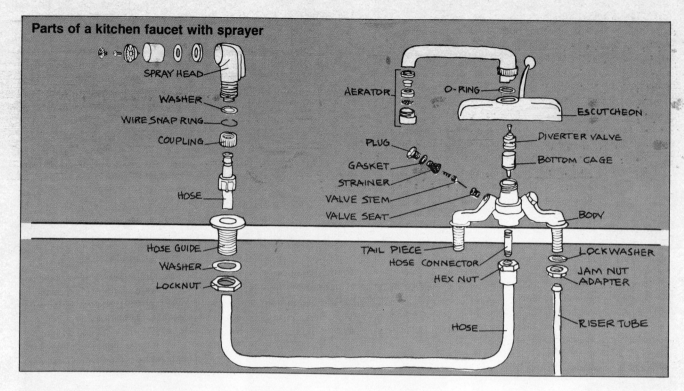

Troubleshooting faucet leaks

Problem	Solution	What it takes
Faucet drips from spout, or forced turnoff is required to prevent dripping.	Replace washer or install basic repair kit or cartridge in washerless faucet. (See pages 17 and 19.)	Household tools, repair parts
Washer-type faucet still drips after above repair.	Dress seat or replace it if worn beyond repair.	Household tools, seat-dressing tool, seat-removal tool, seat-replacement kit
Washer-type faucet drips from around handle.	Repack or install new spindle O-ring. (See page 17.)	Household tools, O-ring or packing
Washerless faucet leaks at handle.	Adjust tension ring, if any. Remove cartridge and repair or replace. (See page 19.)	Household tools, faucet spanner wrench
Kitchen faucet leaks at base of swing spout.	Remove spout-retaining nut, spout and old O-ring. Install new O-ring and replace spout.	Household tools, O-ring

Fixing aerators

After a long period of use, faucet aerators clog up with scale and debris. Instead of replacing the aerator with a new one, try cleaning the old one. Wrap the aerator with tape and thread it out with a pair of pliers moving counter-clockwise. If there was any leakage, replace the old washer with a new one. Turn the aerator upside down and run water through it to clean it. Take the screens out, remembering how they go so you can put them back in the same order. Scrape off any scale with a knife and poke out any blockages with a toothpick. Reassemble and reinstall the aerator.

washerless-faucet parts, when needed, cost more than washer-type ones. These replacements often come in kit form. A complete rebuilding kit contains all wearing parts. Once the new parts are installed, a washerless faucet is ready to begin its new lifetime. It should go far longer than a washer-type before it needs additional repair.

It's a good idea, when repairing washerless faucets, to have a copy of the manufacturer's instruction sheet in front of you. This shows how the faucet is built, and often gives step-by-step instructions for taking the faucet apart and putting it back together. Sometimes the sheet also lists part numbers for replacement parts. Since the directions apply to *that* faucet, they're the ones that should be followed in fixing it. The instructions given here can only apply generally to most kinds of faucets.

If you don't have an instruction sheet for your faucet, you can get one through your plumbing supplier or by writing to the manufacturer. The address, if not available from your dealer, can be looked up in a library. Look in the *A-to-Z Thomas Register*—which is a basic products reference—or in Standard & Poor's *Directory of Corporations*. When you write the company, either give a faucet model number or describe it as completely as possible. This helps the manufacturer send you the correct instruction sheet.

One washerless faucet, of the dual-handled style, is the diaphragm type. A spindle that moves up and down compresses or releases a neoprene diaphragm. The spindle has a swivel end to avoid the twisting friction of a washer and seat. Thus a diaphragm will outlast many washers. The faucet's spindle is kept dry, without packing or an O-ring, by the diaphragm. And so it may be lubricated—and should be—whenever the faucet is taken apart. Use white grease on the spindle's threads to make for smoother on-off action. Diaphragm faucet disassembly is similar to that of washer-type. And repair is usually simple. Replacing the diaphragm with a new one cures most problems. In other words, a single part usually does the trick.

Some newer faucets use metal-to-metal disc construction. While metal-to-metal wear in an automobile engine is avoided through oil lubrication, a disc faucet works well with just water lubrication. The special metal alloys used in the discs are made to resist corrosion as well as abrasion. Some such faucets are guaranteed dripless for ten

Lever-type washerless faucets

Washerless faucets sometimes have only a single control handle, a swinging lever that's attached to the top or rear of the faucet body. A variation is a single, cap-like handle in place of the lever. The lever or cap, when slid or turned from side to side, varies the water temperature. It controls the flow through up-and-down movement. Lever and cap faucets are both valve-type faucets. Both have hot and cold valves inside the body that are actuated by the control handle. Each valve is accessible through hex-head pipe plugs, one on each side of the body. To get at the plugs, remove the faucet trim cover to expose the body. On a kitchen faucet, this comes off after the spout is removed. On a bathroom model, it comes off after handle removal.

Turn off the fixture shutoff or main water supply. Remove the plugs with a wrench—not pliers—and lift out the valve parts. One part is a small strainer screen that keeps debris out of the sensitive valve. If this is clogged, it makes faucet flow sluggish. Clean it, and that may be the total of needed repairs. Screens should be checked periodically for clogging. New valve stems, springs, and strainer screens come with repair kits, but are rarely needed. Install the parts in the same order—and facing in the same direction—as they came out. While you've got the faucet torn down, adjust the handle tightly enough to prevent drift of the temperature setting. A screw may be provided for this.

Taking some cartridge faucets apart can be a baffling task. These function with a washerless cartridge that can be replaced as a unit with a new cartridge if the faucet fails. Their secret is a horseshoe-shaped retaining clip located outside the faucet or inside the handle. To remove the cartridge, you must first get at the clip and remove it.

shaft seals instead of packing. The cap nut is often round rather than hex-shaped. In that case, finger-tight is tight enough. If the nut resists removal, wrap it with tape and try turning it gently with pliers. Counterclockwise removes, clockwise tightens. Spindle-nut removal exposes a knurled, threaded ring beneath, which is also removed by finger power. The valve can then be pulled out past its O-rings. Sometimes a good hard pull is required. Replace the worn discs with a new disc assembly, pushing it in tightly, with the O-rings installed. Screw the ring and spindle-nut back on.

Some disc-type faucets have three rubber seals between their discs. To repair leaks in this type, take the disc assembly apart and replace the seals.

If there's a cap on the handle, look under that. If there's a gap between the handle and spout, look there. And if there's a removable faucet trim panel, take it off to expose the faucet body. Once the retaining clip is off, the cartridge should lift right out of the faucet.

An on-the-blink sink spray-hose can usually be repaired. The spray-hose receives water from a diverter valve in the faucet body. Some sprayers are designed to allow a slight flow of water from the faucet spout even when the spray-hose bottom is depressed. Diverter-valve action will be affected if the aerator and spray-hose head aren't clean and open. Both may be replaced if they cannot be cleaned. Check to see that the hose is free from kinks and sharp bends. If there's any doubt about the hose, remove it (see page 27).

Trouble with the diverter valve is rare, but if the hose isn't the cause, it's most likely the valve. To get at it, remove the faucet spout via its compression nut. The diverter valve rests immediately beneath, inside the center of the faucet body. Lift or screw out the diverter assembly. If it's clogged with foreign matter, it may need replacement. You can probably obtain the parts from your dealer.

One bit of advice when removing faucet trim with pliers: Always protect it by wrapping two layers of masking, adhesive, or plastic electrical tape around it. (If you have a small strap wrench, you can use that instead.) Finished surfaces of faucets, including their escutcheons, may scratch easily. These scratches don't only look bad; they can make the faucet rust and corrode.

Any faucet that cannot be repaired, or is so badly deteriorated that it is not worth repairing, can be replaced. See pages 24–25 for replacement instructions.

When a faucet in the wall drove me up it

I know tub and shower faucets operate pretty much like sink and lavatory faucets. So when the shower head began to leak—no problem. Just take the valve apart and replace the washer or dress the seat like any other washer-type faucet. The hooker was that none of my wrenches could get at the valve bonnet nut to turn it—it was recessed too deep in the wall. Talk about frustration!

Getting mad won't help. Beg, borrow or steal a plumber's socket wrench (see page 13) to remove a washer-type spindle recessed in a wall. To find the wrench size needed, bend a pipe cleaner in two so that its tips touch opposing faces of the nut, and measure the distance between tips. Or, if the cleaner's too short, cut a length of coat-hanger wire and bend it so that its tips can gauge the nut's size. Use a plumber's socket wrench which fits that size snugly.

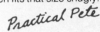

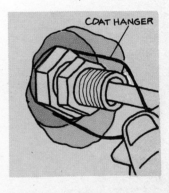

Cleaning a diverter valve

About valves

Valve problems

Globe valves can have all the problems that beset washer-type faucets. But, most often they suffer from leaking around the shafts due to deterioration of the packing, freezing, and scale.

Any inner scale deposits render globe valves (and gate valves, as well) incapable of being securely turned off. Complete disassembly and cleaning is called for.

If a pipe that contains a valve freezes, it will damage the valve—either by cracking the body or ruining the washer and spindle. Replacement spindles can sometimes be found, but there's no cure for a cracked valve body other than a new valve.

Ground-key valves, after much use, may suffer from metal-to-metal wear between the key and valve body. There is no cure short of replacement.

Parts you can get

Replacement parts for valves are not widely available. You *can* get replacement washers, however, as well as packing for leaking spindles. (Installation parallels that for a faucet.) Worn-out gate valves can be rebuilt, but that's a shop job.

Valves are merely mid-pipe faucets. The best ones are made of brass. Others are made of steel or plastic. All come ready for attachment to the kind of pipe they are designed for. Some can be adapted to use with other pipes also.

A **globe valve** is built very much like a washer-type faucet. It has washer, spindle, packing, and seat. With this type of valve, water flow is somewhat restricted due to the path it must take (see illustration below). If this drawback didn't exist, no other kind of water valve would be needed.

Highly useful for some systems is the **stop-and-waste valve**. Also called a stop-and-drain valve, it's a globe valve with an added feature: an auxiliary valve that opens into the closed end of the main valve. With the main valve off, the auxiliary valve can be opened to release water in the line beyond the valve. This lets you easily drain the pipes when the house is to be vacant and unheated in below-freezing weather (see page 90). Some types have a side drain-screw rather than a valve.

All globe valves should be installed with water-flow direction in mind. Many globe valves have directional arrows on the valve body. If you are using one that doesn't, install the valve so that water flows in *below* the valve seat, through it, and then out around the valve spindle. If one is installed backward, water pressure would be on the spindle, which might tend to leak, even with the valve turned off. Directional alignment applies only to water flow, not to the angle of the valve handle. This may be pointed in any direction: up, down, or sideways. Position it for convenience.

Where full water flow is needed, a **gate valve** is used. A gate valve features a wedge-shaped gate or stopper that presses firmly into a reciprocal chamber when the valve is closed (see illustration below). The two surfaces meet tightly, sealing off flow. When the valve is opened, the gate lifts completely out of the chamber, permitting full water flow through the valve. Moreover, flow through a gate valve is straight. The water does not have to go around a corner, as it does inside a globe valve. This, too, works for full flow. The careful parts fitting and large size of a gate valve make it costly—considerably more so than a globe valve. Therefore, gate valves are not used unless full flow is required. They're normally used in the service entrance only.

Another kind of valve—the **ground key**—is used for water, gasoline, and solvent liquids. It is also used for a buried, outdoor service-entrance line. A cone-shaped key is held tightly in a cone-shaped body. The mating surfaces are machined smooth so that no leakage can take place between them. A handle attached to the key is used to turn it within the body. Both key and body have a round hole through them.

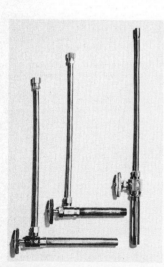

Water supply should be controlled with fixture shut-off valves below each fixture, with one for hot-pipe line and one for cold-pipe line below sinks or lavatories.

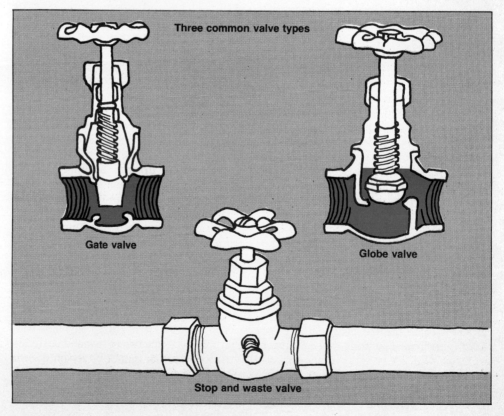

Three common valve types: Gate valve, Globe valve, Stop and waste valve

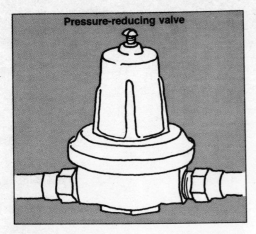
Pressure-reducing valve

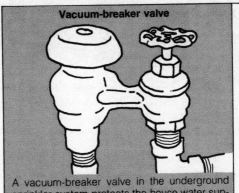
A vacuum-breaker valve in the underground sprinkler system protects the house water supply in case of back-siphonage.

How vacuum breaker works. If the water starts to flow backward, suction pulls the disc down, closing off the opening.

When the handle is turned and the holes line up, flow occurs; when they are only partially aligned, partial flow occurs. When they don't line up, no flow occurs. They are sometimes built with a tab or recessed key so that a wrench or special tool must be used to operate them. In this way, the water utility uses them in underground service pipes to homes to turn off water to customers who don't pay their water bills.

Houses built in low areas may suffer from excessively high water pressure. When the pressure reaches more than 80 psi, it puts a strain on faucets and valves, including the toilet-tank float valve. High pressure also wastes water because it makes too much flow out of the taps. Finally, it makes for noisy plumbing. There's an easy cure: installation of a **pressure-reducing valve.** Even if the water company will not install one in its water main, you can put one in your house. It not only lowers house water-pressure levels; it evens out any fluctuations in pressure.

Vacuum-breaker valves serve a special purpose (see margin at right). They are used with underground sprinklers and tankless flush-valve toilets. The best quality tank-type toilets—recognizable by an air vent reaching above the tank's water line—also have vacuum-breaker valves. Under normal conditions, a vacuum-breaker valve is inactive. But should a vacuum occur in the water-supply system, a spring-loaded valve inside the vacuum breaker opens up. This lets air into the valve, creating an air gap. Since a supply-line vacuum cannot back-siphon water through the air gap, your water system is protected from contamination by any polluted water source beyond the valve.

Other useful valves include **fixture shutoffs.** These are usually located immediately beneath a fixture, either on the wall or floor. Usually a fixture's water comes to it from the wall. Then **angle-stop shutoffs** are used. When a fixture's water supply comes through the floor, **straight-stop shutoffs** are used. Whatever the type, in case of emergency, both shutoffs should be accessible from the room in which the fixture is located. Those in a branch supply-line below the floor are less handy. Adding accessible shutoffs to a fixture without them is a worthwhile one-day plumbing project. For fixture hookup with riser tubes and shutoffs, refer to pages 10, 25, and 28.

Valve placement in a water-supply system begins at the buried service-entrance line out front. Between the water main and the house there may be a ground-key valve. This is where the water company usually controls your service from. Another valve—a gate valve or a large stop-and-waste valve—should be installed next to the meter, which is located close to the point where the water service enters the house—usually indoors in freezing climates, outdoors in mild climates. This becomes your emergency shutoff valve, stopping all water flow to the house.

Other gate valves are placed below both inlet and outlet connections to a water softener. A third valve is placed between the tees to allow water to flow with the softener disconnected. Next in the system is the cold-water inlet valve for the water heater. If the inlet has a ¾-inch diameter, a globe valve may be used. If the diameter is ½ inch, a gate valve is necessary.

Sometimes intermediate valves are used—for example, a stop-and-waste valve on the basement or crawl-space end of a supply pipe leading underground to a garage, or on hot or cold branches leading to a distant bathroom.

Finally come the valves for branches to fixtures. Individual hot and cold shutoffs for each sink or lavatory fixture are best.

For homes with hot-water heat, one more valve should be installed. This is a backflow-preventer valve, which goes between the water supply and the hot-water boiler. Filling the boiler is its main job, but this valve doubles as prevention against back-siphonage. Without it, boiler water could enter the potable water system.

Cross connections and vacuum-breaker valves

A cross connection between a possibly polluted water source and your potable house water can cause sickness and even death. Any submerged water outlet is a cross connection. A garden hose with its end left in a wading pool is a cross connection. If an upstairs faucet is turned on, water from the pool can flow backwards through the open hose bibb and into the potable water system. Backflow of bacteria can even occur through a closed faucet or valve. A broken fire hydrant down the block can create back-siphonage in all houses on the street, including yours. As shown, an underground sprinkler head is a potential cross connection. All underground sprinkling systems should have vacuum-breaker valves. These, in the event of back-siphonage, create an internal air gap that keeps polluted water out of potable water. A toilet tank should have a vacuum-breaker or anti-siphon intake valve. No fixture faucets should discharge below the rim of the basin. If any do, replace them with higher faucets. Don't leave the ends of hoses resting below the flood rims of pools, pails, sinks, or ponds.

FAUCETS 23

3. SINKS AND BATHS

Installing faucets

If you have repaired a leaking faucet too many times—or if it has gotten unattractive with age—you may want to replace it with a modern one. The steps at the right show how it's done.

If you are not planning to replace the riser tubes beneath the faucet, the job goes best if the new faucet has exactly the same tailpiece as the old one. This governs how you connect the fixture's water supply to the faucet. The handiest faucets are the "do it yourself" type. They come with ½-inch threaded tailpiece adapters that accept the bullet-nosed ends of riser tubes. A ⅜-inch fixture flange nut (sometimes called a jam nut or supply nut) holds things together. This type of faucet will also fit a ½-inch pipe coupling or take a ½-inch flare adapter. Making faucet water-supply connections with riser tubes is recommended. It saves a great deal of extra effort.

Faucets with other kinds of tailpieces can be adapted to risers, but not quite as easily. Some lavatory faucets end in plain-ended, ¼-inch O.D. (outside dimension) copper tubes, which you have to connect to the water supply. A riser-tube hookup still is best. Cut off the upper end of the riser tube, flare it and the faucet tailpiece, and connect the two with a ⅜x¼-inch reducing flare coupling and two flare nuts.

Still other faucets end in tailpieces with ¼-inch pipe threads. These are about 3 inches long and are held to the faucet with fixture flange nuts. They are most common on kitchen-sink faucets. Discard both tailpieces and make the connections with bullet-nosed riser tubes, using the same fixture flange nuts.

Some kitchen faucets merely end in a threaded opening, but not tapered pipe threads; these accept fixture nuts. If the opening is too large to accommodate a standard riser tube, get a rubber bullet-nosed fixture washer to put between them. Then a larger fixture nut can be used to secure the connection.

If you're in doubt as to how to adapt a new faucet to the water supply, tell your dealer what your fixture's water-supply setup is. Ask him to sell you the parts needed to make the new hookup—with riser tubes. (In nearly every case, a pair of riser tubes does it best.) At the lower end, these tubes must fit into a compression nut at the wall or floor, and if one is not already there, you'll have to provide it. A new pair of fixture shutoffs provide compression. Next best but lower in cost would be a pair of compression adapters for fixture supply. Use *straight* for a floor installation, *angled* for a wall installation. The lower end of the riser tube slips into the compression fitting and tightens to form a watertight, yet removable, coupling. Compression couplings use a brass ferrule and compression nut that squeezes the ferrule onto the riser tube for a snug, watertight connection.

Modernizing fixtures

You can change a bathroom or kitchen sink at the same time you change faucets. Or you can change the fixture and retain the old faucet. How you do the job, as well as the existing parts you'll want to save, depends on what you have now and what you're replacing. For example, an in-the-counter sink or lavatory bowl can be reused in a new counter-top installation. Merely remove it from the old counter and mount it in the new one. Or you can discard it altogether. A wall-hung lavatory or sink most likely would be junked and replaced with a cabinet or counter-top unit. You can make the cabinet yourself or buy one. You can have cabinets made to order or purchase component units combining them to fit your space. You can also buy prebuilt units, completely finished and ready for bowl installation. Standard single-bowl lavatory units are 17 inches deep, 30 inches high, and available in lengths ranging from 30 inches to 4 feet. Kitchen-sink cabinets are single or double-bowl and vary in length from 3½ to 8 feet. Most have 2-foot-deep counters and stand 3 feet high. In all cabinets, space is provided for water-supply and waste hookups. The selection is enormous; your dealer will be glad to show you what's available.

Counter sink and lavatory bowls are held from underneath by levers and bolts arranged around the perimeter of the bowl. A stainless-steel flange, contoured to fit the bowl, surrounds it. T-shaped in cross-section, the flange is forced down tightly against the counter top at the same time the bowl is pressed tightly up against the flange. Levers placed in the flange and adjusted with bolts do both jobs as the clamps are tightened from below. Plumber's putty placed between the bowl and flange and counter top and flange prevents leaks. Don't use ordinary putty. It

---

What it takes

Approximate time: Replacing a faucet is a one-morning or afternoon project. Changing a lavatory or sink fixture takes most of a day.

Tools and materials: A pair of adjustable open-end wrenches and a basin wrench will handle the faucet change. For fixture modernization, you'll need most of the tools listed on page 26, the new parts for installation, plumber's putty and putty knife, and new riser tubes.

Fittings

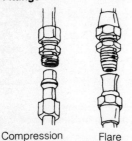

Compression   Flare

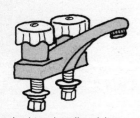

Standard two-handle mixing faucet

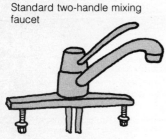

Single-handle mixing faucet

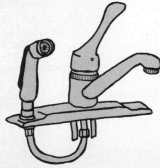

Single-handle kitchen faucet with spray attachment

24 SINKS AND BATHS

stains and becomes brittle. Use only plumber's putty made for this purpose.

The stainless-steel flange and clamps to go with it come with the bowl. Mount the faucet in the fixture before you put the fixture in place. It wouldn't hurt to have the upper riser connections completed; that will save you from making the connection while lying on your back under the fixture. Slide the new fixture to the wall and fasten it. Then, make the lower riser-tube connections, hook up the trap as explained on pages 26 and 27, and turn on the water. Your new faucet is ready to go.

Replacing faucets

1. To install a new lavatory or sink faucet, turn off cold- and hot-water supply valves and turn faucet on to drain the pipe. Loosen hold-down nuts, using a basin wrench, if needed, to get at them.

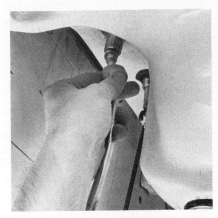

2. To remove the old faucet, take off the hold-down nuts and washers and remove piping attached to the faucet tailpieces. If riser tubes have been used, remove them from their compression fittings.

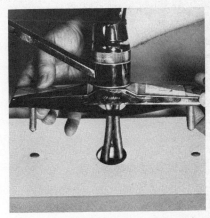

3. The new faucet should match the old faucet's hole spacing, with two holes 4 inches apart for lavatories, or 8 inches apart for sinks. There may also be a center hole for mixing faucets, and a fourth hole for a sink spray attachment.

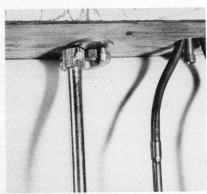

4. If faucet has no gasket, seat it with plumber's putty between it and the sink top. Attach washers and hold-down nuts firmly (but not tight enough to crack a china fixture, if that's what sink is).

5. Connect the riser tubes or pipes from the water supply to the faucet tailpieces. After finger-tightening, use two wrenches—one to hold and one to turn—to keep from twisting the tailpieces.

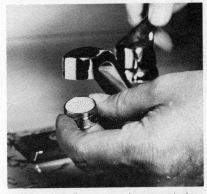

6. With water lines securely connected, remove aerator if faucet has one. Then turn on both hot and cold water at full pressure for one minute to clear pipe scale or other debris from system. Shut off water and reinstall the aerator on the faucet.

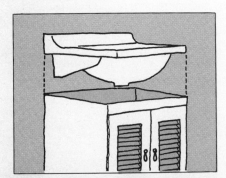

Old pedestal-type lavatories can be replaced with modern units that come complete with cabinet to match almost any decor. Or you can build a cabinet to go under a wall-hung lavatory basin.

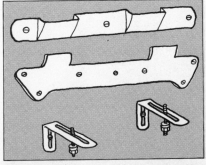

Various types of brackets are available for hanging wall-mounted lavatories, using either toggle bolts or, better still, a wall backing board. Front support legs are available on some models.

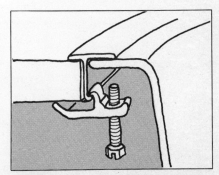

Levers placed in the sink flange hold counter-top units in place. Tightening the bolts draws the flange tightly against the putty seal between the flange and sink or counter top. The flange, levers, and bolts are supplied with the fixture.

Installing traps

What it takes

Approximate time: Under an hour per trap.

Tools and materials: Monkey wrench or trap-and-fixture wrench. Possibly a hacksaw. A new trap to fit. Use P-traps for waste pipes coming from walls; S-traps for pipes from floors. Double-bowl kitchen sinks may require other new parts if the two bowl drains are to be connected to a single trap. The trap may go on the right, left, or be centered between the two bowls. (Note: A centered trap calls for centered drain fittings.) You'll also need an additional slip nut to connect the trap to the adapter. If you're joining a 1¼-inch lavatory trap to a 1½-inch trap adapter, use a 1¼-by-1½-inch reducing slip nut.

Planning hints: Choose a time when the fixture will be unused for an hour or so.

Drain traps for sinks, lavatories, tubs, or showers are usually either the P-type or the S-type shown below. On some models, known as fixed traps, the J-bend and trap arm are one piece. More common are traps with a removable J-bend that can be swiveled to one side to remove tailpiece connections. Cleanout plugs make unclogging easier; some traps don't have them.

Replacing rotted O-ring washers or corroded metal slip nuts or pipe sections calls for disassembly. Shut off water supply pipes, place a pail or plastic wastebasket under the trap, and remove cleanout plug to drain water from the trap. Loosen slip nuts with wrenches, one to hold and one to turn with, so that you don't twist the fixture or waste pipe parts the trap is attached to. If a slip nut won't budge, saw through it gently with a hacksaw until it expands enough to come off easily. But don't saw deeply enough to cut into the threads of the trap arm or fixture tailpiece.

Lavatories usually use 1¼-inch traps; sinks and laundry tubs, 1½-inch traps. It's wise to replace an S-trap with an S-trap, or a P-trap with a P-trap the same size. The drum traps for tubs (page 37) found under some old bathroom floors may need to be replaced with a P-trap if no drum trap replacements are available where you live. Consult your plumbing supplier on this.

To install the new trap (or the old trap with its new washers and slip nuts), attach the washer and one end of the J-bend to the tailpiece with a slip nut, threading it on just enough to hold these pieces together. Slide two slip nuts onto the trap arm, with their threads facing toward the pipes to which they will attach at each end. Slide washers onto both ends of the trap arm. Push the trap arm into the drain stub-out (or trap adapter, if there is one), sliding it to line up the J-bend and trap-arm connection. When everything lines up without straining, connect the washer and slip nuts and tighten all the slip nuts with wrenches. But don't tighten so hard that you bend the trap.

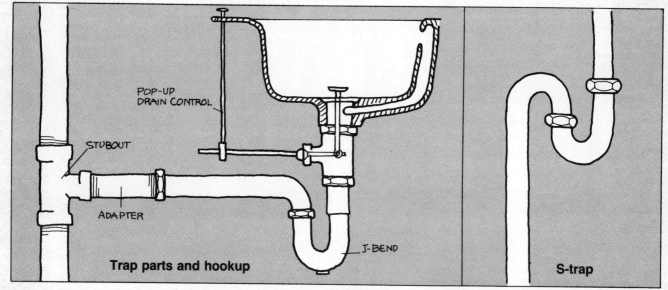

Trap parts and hookup — **S-trap**

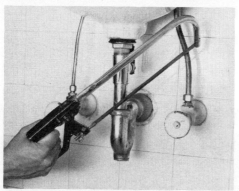

1. To replace a corroded trap, remove the old one without damaging the fixture or trap adapter at the wall or floor. To make the job easier, put penetrating oil on the thread slip nut the night before. You can even saw through a trap, if necessary, to get it out.

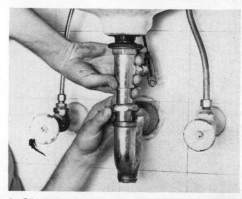

2. Slip the new trap arm into the adapter and put the J-bend up onto the fixture's drain tailpiece. Use new slip nuts and washers installing the nuts facing the threads they attach to. Before tightening with a wrench, tape slip nut to protect it.

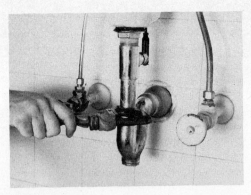

3. The last step is to tighten all the slip nuts to secure the trap in place and make it free of leaks. To further guard against leaks, you can precoat the three trap connections with silicone-rubber sealant on metal, PVC, and polypropylene-plastic traps. Don't use it on black ABS plastic.

Replacing a sink sprayer

If you can't repair a sink sprayer, you can easily replace it. Loosen the coupling nut holding the hose end to the faucet underneath the sink. (If connection is made with tapered pipe threads, simply twist the hose counterclockwise to free it.) Stand aside as you disconnect as some water in hose may drain out. Pull the hose out through the sink top and take it to your plumbing supplier. Have him match it with an equally long replacement; if it doesn't match your faucet tailpiece connection, he can provide an adapter.

Upgrading a shower head

If your old shower is a tired performer with a shabby look, you can update it with one of the new shower heads shown at right. Most simply thread onto the old shower arm. Remove the old shower head by turning it counterclockwise with a wrench. Put pipe thread dope on the shower arm threads—or else wrap threads with TFE Teflon tape and thread on the new head and tighten with a wrench.

If you want to save water—and the cost of heating it—get a shower head cutoff valve or a flow-restricting device from your supplier. Either attaches onto the shower arm before the shower head is attached (see drawing below). The flow-restricting water saver cuts all water flow to 2 gallons per minute (versus the 5 gpm many showers now use). The cutoff valve is controlled by the user manually.

Stopgaps and copouts

Taping a trap: If replacing a leaking trap is inconvenient, you can postpone the job for a while by using plastic electrical tape. Dry the outside of the trap, then make several wraps with the tape. Such a stopgap may last for months. But don't push your luck. Get a new trap and make the replacement as soon as you can.

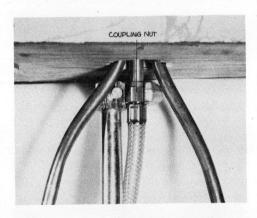

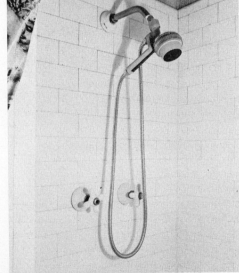

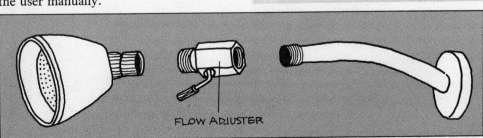

The mark of an amateur
Boy, have I got trouble. I just installed a beautiful new chrome shower arm. At least it *was* beautiful. Trouble is I scarred it up so badly with the wrench that it looks worse than the old one did. A professional plumber told me that when he needs to tighten exposed plumbing pipes he gives them a double wrap of plastic electrical tape to protect them from wrench teeth. Afterward, the chewed-up tape strips off, leaving the pipe undamaged. I'm going to try that next time. I may even buy a strap wrench for tightening exposed pipes. It sure beats scarring a smooth finish with a pipe wrench.

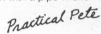

Fixing shower heads
Water holes in a shower head's faceplate or around its rim may clog. Remove screws or knob holding faceplate, soak overnight in vinegar and scrub with a brush. Poke out any blockages remaining with a toothpick. If shower head drips or pivots stiffly, unscrew head and collar connected to shower arm, replace washer, and smear petroleum jelly on swivel ball before reassembling.

Installing fixture shutoffs

Today's houses are usually built with fixture shutoff valves beneath every sink, lavatory, and toilet. The valves let you change a faucet washer, repair a fixture, or stop a flooding toilet without shutting off the water supply for the entire house. If your house doesn't have fixture shutoffs, you may want to add them.

All piping between the water-supply pipe coming through the wall or floor and the fixture tailpiece can be removed and replaced with a fixture shutoff valve and a riser tube. Follow the sequence of steps shown in the pictures below. Get the angle shutoff valve shown for water-supply pipes coming through the wall, or a straight shutoff valve for supply pipes coming through the floor.

Modern shutoff valves end in ⅜-inch or ½-inch compression couplings that fit ⅜- or ½-inch riser tubes. The end of the valve that will be connected to the wall or floor water-supply pipe may be made to fit threaded pipe, sweat-soldered copper tube, or plastic tube. Take the old pipe you have removed from the water-supply pipe to your plumbing supplier so that he can provide a valve that will fit it, or an adapter that will let you connect the shutoff valve to the water-supply pipe.

What it takes

Approximate time: Give yourself an hour and a half for each valve. If riser tubes are already in place, the job will go more quickly.

Tools and materials: Monkey or open-end wrenches, pipe dope or TFE tape; propane torch, soldering flux, and 50/50 wire solder for copper water-supply pipes. Possibly a pair of pipe wrenches for removing threaded piping.

Planning hints: In most cases, much of the house will be without water when you make valve installations. So tackle this work when others are not going to be using water. Know beforehand what adapters and fittings you'll need and have them on hand. At a minimum, this includes the necessary fixture shutoff valves (angle-type for wall supplies, straight-type for floor supplies). Sinks and lavatories take two valves—one for hot water, one for cold. Toilets take one. If you use new riser tubes, get the bullet-nosed type for sinks and lavatories and the flat-ended type for toilets. All risers must be long enough to reach from the valve to the fixture's tailpiece. If they aren't already present, you'll need enough fixture flange nuts to connect all riser tubes to the faucet tailpieces or toilet valve.

1. To install fixture shutoff valves, turn off water supply, unscrew nuts connecting old pipes to fixture tailpieces, and remove piping connected to the water supply stub-outs. If there isn't enough give to remove the old pipes, hacksaw a small section of the old pipe just above the joint that connects to the stub-out.

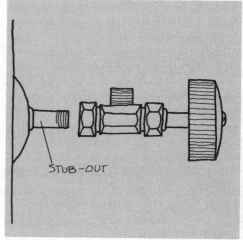

2. Unscrew elbow or other piping connected to the stub-out (or melt apart soldered connections) and get a fixture shutoff valve that will attach to your stub-out. Put Teflon tape or pipe dope (joint compound) on the threads and attach valve to stub-out so that its compression nut lines up vertically with the faucet tailpiece.

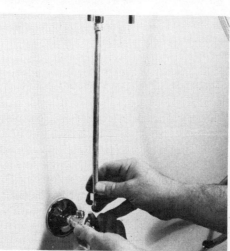

3. Cut a riser tube to fit between the new shutoff valve and faucet tailpiece with its lower end inside the compression nut on the valve but not quite touching the shoulder. Allow for some bending if needed to get a lineup between valve and tailpiece.

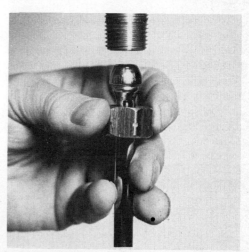

4. Fit lower end of riser tube into compression nut on valve and tighten the nut. Fit the upper end to the faucet tailpiece, bending as needed, and tighten this connection. When all nuts are secure, remove aerator (if faucet has one), turn on water, and test as described in step 6, page 25.

28 SINKS AND BATHS

Caulking

A tub-to-wall crack looks ugly and lets water leak against the wall, damaging it. The remedy is to rake out all the old caulk and reseal the joint with silicone tub caulk, which lasts more than 20 years outdoors and even longer indoors.

For the patch to work, the gap between the tub and wall must be wide enough for a decent-sized bead of caulk, and its depth should be roughly the same as its width. If your tub-wall joint is deeper, fill it in with rope oakum (jute fibers coated with preservative) or closed-cell plastic foam to a depth of about ¼ inch before you caulk.

If the wall has been damaged by water, you may have to replace it with one of the waterproof materials made for bathroom walls. If you do, leave a ¼-inch gap between the tub and wall material, even if it's tile.

Silicone tub caulk tends to bulge in spots and dip in others as you apply it. Apply masking tape above and below the tub/wall gap and have dampened cloths handy to wipe off hands and spills. Hold spout at a 45-degree angle and parallel to the joint and move it forward as you squeeze the bead in the joint.

A bathtub has two ends and two long edges, including the one at the floor. Do the end away from the faucet first, because it is looked at least. Apply caulk to the entire end joint. Try for neatness but not perfection. Fill the joint with caulk, then immediately tool it down—before any skin-over, or surface-hardening, can occur. Use a finger, the curved handle of a spoon, an ice-cream stick, or a tongue depressor to put a concave surface on the bead of caulk. Wetting the tool will keep fresh caulk from sticking to it.

When you have the first tub/wall section tooled into shape, go on to the next section. If the bead gets worse instead of better as you tool, stop. The more you try to smooth skinning-over silicone caulk, the rougher it is likely to get.

The use of toxic silicone caulk solvents is not recommended, especially in an unventilated bathroom or bathtub enclosure. If the smell of caulk seems strong, open a bathroom window for ventilation. Also, if you have a skin allergy, better not touch the caulk. Use rubber gloves.

A day later, you can repair any serious roughness by slicing excess caulk with a single-edged razor blade. At the same time, you can scrape off any unwiped caulk spills. Since the silicone bead cures clear through, this is like cutting soft rubber, which is what the bead is. Do not try to paint silicone tube caulking, however. It comes in colors so you won't need to.

Don't mess around
Have trouble caulking without getting all messed up? I did too until I tried using masking tape. Lines of tape laid along both sides of the tub-wall crack lets the bead of caulk go only where you want it. The bead can be tooled after applying, still without messing you up. Finally, when the caulk has gotten a little "tack" to it, you can pull off both strips of tape, leaving a nice, neat bead. Hey, and no mess.
Practical Pete

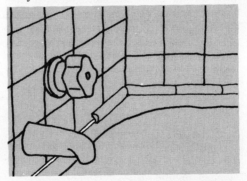

1. Apply masking tape above and below gap between tub and wall. Lay a bead of caulk in the gap, pushing tube forward as you go so that spout forces caulk into the gap.

2. If the caulk won't hold or you think it is unsightly, you can cover the gap with a quarter-round ceramic tile, which is available in kits from plumbing supply houses.

What tub caulk to use in the bathroom

Caulk type	How sold	Advantages/disadvantages	Solvent	Colors	Relative cost
PVA tub-and-tile	Tubes, cartridges	Easy to use. Excellent adhesion, but shrinks. Should be painted. Short-lived.	Water	White	Lowest
Acrylic latex tub-and-tile	Tubes, cartridges	Low cost, easy to use. Excellent adhesion. Cheapest type recommended. Should be painted.	Water	White	Low
Silicone rubber tub-and-tile	Tubes	The best. Lasts and lasts. Accepts great joint movement. Never becomes brittle. Needs no painting, but hard to use neatly.	Paint thinner, naphtha, toluol, xylol	White, blue, pink, green yellow, beige, gold	Highest

SINKS AND BATHS

4. TOILETS

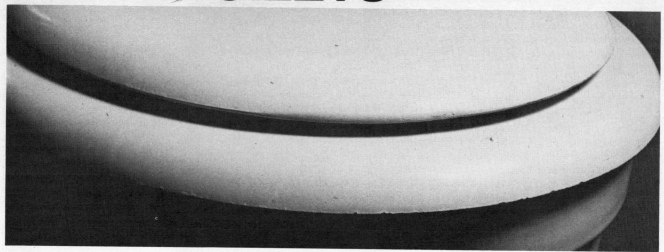

Unveiled: the mystery of the flush toilet

The flush toilet was invented more than 300 years ago but many people are still baffled by its inner workings. When a toilet won't flush, doesn't fill or won't stop running, they jiggle the handle and then call a plumber. If they understood how a toilet mechanism works, they could repair it themselves.

What happens is this. Like the "hand-bone-connected-to-the-wrist-bone" song, the handle connects to the trip lever which in turn connects to the tank ball, as shown below. Pressing the handle raises the tank ball from its seat, letting water rush from the tank into the toilet bowl (as shown at left, top illustration).

As water level drops, the float lowers, opening the intake valve to let water enter the tank (left, center). Releasing the handle lets the tank ball drop into its seat, closing off the tank and allowing it to fill.

As the tank fills, the float rises until it closes the intake valve when the water reaches its fill-level (left, bottom). Meanwhile, the emptied toilet bowl is partially filled by water flowing from the refill tube into the overflow tube as the tank is filling.

What to do when a toilet mechanism misbehaves, uses too much water, or needs replacement is explained on pages 31-35. For the special problems of unclogging blocked toilet drains, see page 38.

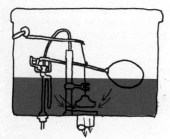

Water rushes from tank, causing

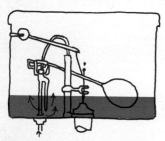

float to drop and open intake valve,

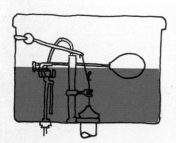

filling tank with fresh water and closing valve.

Parts of a flush toilet

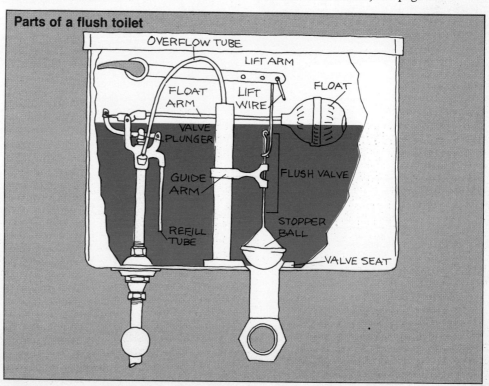

Troubleshooting toilet tanks

Problem	Solutions	What it takes
A. Tank fills but water keeps flowing. See steps 1-6 below.	Bend float rod to lower ball. Replace leaking float ball. Replace washer and seat in intake valve, or install a new valve.	One minute to an hour using inexpensive replacement parts and household tools.
B. Tank doesn't fill but water keeps flowing. See page 32.	Make sure lift wires and guide arm are not binding so tank ball can't seat firmly in flush valve seat. Check trip lever and trip handle for binding. Bend or replace parts and rotate guide arm on overflow tube as needed.	Less than an hour using household tools and inexpensive replacement parts.
C. Water level is too low or too high. See page 32.	Bend float rod up or down to raise or lower water level.	A minute or so.
D. Tank doesn't flush completely. See page 32.	Rehook upper lift wire through alternate hole or holes in trip lever. If that doesn't work, shorten wire and then rehook.	A few minutes. Tin snips and needle-nose pliers may be needed.
E. Splashing noise from tank during filling. See page 32.	Make sure refill tube feeds directly into overflow tube, bending if needed. If that doesn't solve it, replace seal and washer in intake valve, or install a new valve.	Less than a minute to a half hour if needed parts can be obtained readily.
F. Air condenses on tank, making it sweat. See page 32.	Line tank with plastic foam sheets.	Less than half an hour if material is readily available.

Problem A: Tank fills, but water keeps flowing.

1. If water shuts off when you lift the float rod, gently bend the rod down until the float rests about ½ inch lower in the water. Water should now shut off when its level is about ½ inch below the top of the overflow pipe.

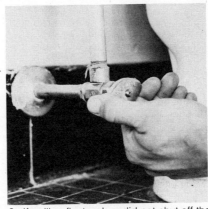

2. If pulling float rod up did not shut off the water, turn off the shutoff valve below the tank and flush the toilet to empty it. If there's no shutoff valve below the tank, look for a stop valve on supply line feeding the toilet, or shut off main valve.

3. If more than half of float is under water, it may have a leak. Unscrew it from the float rod and shake it to see if there's water inside. If there is, replace it. If there isn't, reinstall it.

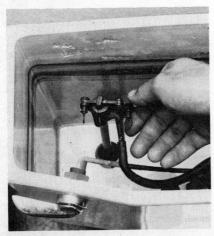

4. If toilet still runs on after filling, remove the screws that hold the float rod and its attached linkage to the intake valve. These screws are usually thumbscrews or have L-shaped heads you can hand-turn.

5. Remove float rod and its attached linkage from the intake valve and pull the valve plunger up out of its seat. You may need to pry it up gently to get it started moving out.

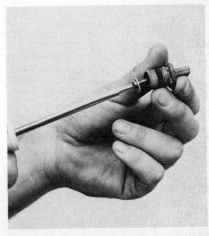

6. Replace the washer at the base of the plunger and the second washer or packing that fits in a groove partway up the plunger's body. If toilet still runs on after filling, install a new fluid level control valve (page 33).

TOILETS

Problem B: Tank doesn't fill

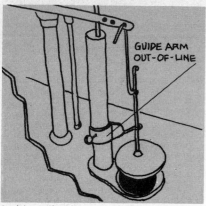

1. A bent lift wire or an out-of-line guide arm may keep tank ball from dropping into its seat to close the tank. Straighten lift wires and rotate guide arm on overflow tube until tank ball slides down easily and snugly into its seat. Tighten clamp on guide arm to prevent slippage. If ball is worn, replace it.

2. If trip lever stays up after flushing, try bending and oiling it, and then tightening the screw that fixes the trip lever to the toilet handle. If the trip lever still won't release, replace it.

Problem C: Wrong water level

1. The tank water level should be no higher than the level line marked inside the tank—or ½ inch below the top of the overflow tube. To save water, you can lower the level as far as it will go and still flush adequately. To lower water level, gently bend float rod down. To raise level, bend it up. Flush and test after each adjustment.

Problem D: Incomplete flushing

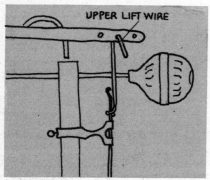

1. A partial flush is caused by the tank ball not being lifted high enough when the lever is tripped. To correct this, first make sure that the lower lift wire is screwed all the way into the tank ball. If it is, unhook the upper lift wire from the trip lever and hook it in a hole position closer to the valve. If that doesn't solve the problem, bend the hook again, making the trip wire shorter.

Problem E: Splashing

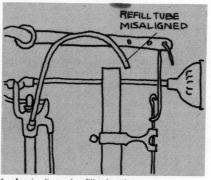

1. A misaligned refill tube that spouts water directly into the tank will cause splashing, often accompanied by a lack of trap-sealing water in the toilet bowl. Reposition refill tube so that it spouts into the top of the overflow tube. Do not let the tube's end reach below the tank water level. That would make it siphon tank water away, causing constant slow running of water.

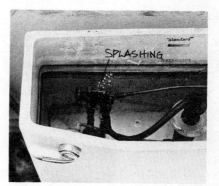

2. A faulty toilet inlet valve is rare, but can cause splashing. If the valve is at fault, you will be able to see it leaking during tank refill. Either replace the entire valve assembly or repair the valve as shown on page 31. For either job, water supply to the toilet must be turned off.

Problem F: Tank sweats

1. In a humid climate when house water is cold, the toilet tank may sweat. Installing a tempering valve that mixes hot and cold water to refill the tank might stop sweating, but that uses scarce fuel to heat water. Try insulating the inside of the tank. Kits are available. Start by turning off the water and completely drying the inside of the tank. Use a sponge and paper towels, or a hair dryer.

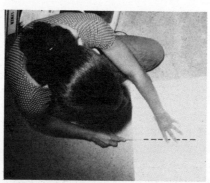

2. Cut a piece of foam-board to fit the inside of your toilet tank. Bottom, sides, front, and back should be protected up to the highest water level. Some of the pieces may have to be sectioned to fit around tank hardware. Try not to leave any gaps.

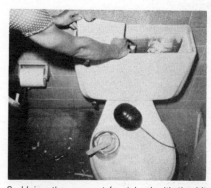

3. Using the cement furnished with the kit, glue each piece into place. Scraps and cement can be used to fill any openings. The insulation's thickness reduces the volume inside the tank, also reducing the amount of water used for each flush. When the cement has set, the tank's ready for use.

Flush valve parts

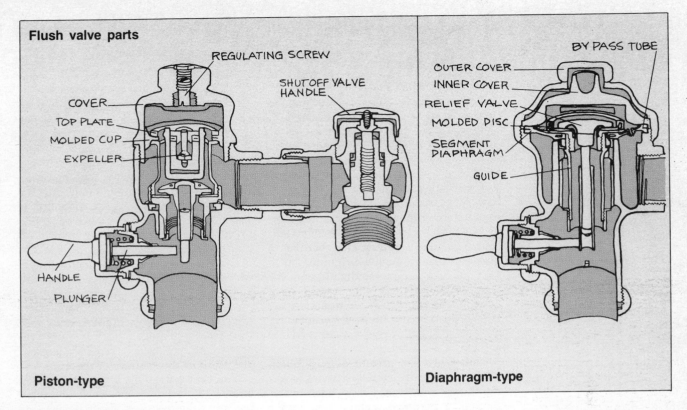

Piston-type | Diaphragm-type

Tankless flush valves

Tankless toilets with pressure valves—like public toilets—seldom act up. When they do, repairs are easy. How you go about it depends on whether your toilet has a diaphragm-type valve or a piston-type. If there's an adjusting screw and nut on the valve's cover, it's a piston type. A diaphragm type has no cover adjusting screw, and its inner cover is much wider than the pipe leading to it. Either type has a shutoff valve located in the pipe that comes from the wall. It may be controlled by a knurled handle or a screw you can reach after a cover nut is removed (see above). To shut off water when making repairs described in the chart below, turn handle or screw clockwise. When repairs are complete, open shutoff valve only one or two turns on a diaphragm type. Open valve fully on a piston type, whose flow is then controlled by adjusting the regulating screw and nut on the pressure valve cover.

Troubleshooting tankless flush valves

Problem	Solution	What it takes
Incomplete flush	Turn regulating screw beneath cover to the left to increase flush duration (piston-type only). In diaphragm-type, molded disc and diaphragm not assembled tight. Take valve apart and tighten disc by hand.	Strap wrench or pipe wrench and tape, screwdriver.
Too much water during a flush	As above for piston-type unit, but turn screw to the right. No regulation required on diaphragm-type unit.	Strap wrench or pipe wrench and tape, screwdriver.
Water runs after flush	Relief valve or piston not seated properly, or foreign material clogging bypass tube. Disassemble working parts of valve, rinse off, and replace.	Strap wrench or pipe wrench and tape, screwdriver.
Short-flushing	On piston-type only: (1) Water bypassing molded cup. Flare out or replace cup to fit tighter in valve cover. (2) Loose piston assembly. Tighten screws on top plate. Screw guide hand-tight. (3) Bypass in top plate enlarged from corrosive water. Replace top plate and expeller.	Strap wrench or pipe wrench and tape, screwdriver.

New wrinkles

One way to solve persistent flushing problems is to replace tank ball, lift wires, and guide arm with a new rubber flapper. Like the tank ball, it rests in the flush valve seat but is raised by a chain connected to the trip lever.

Noisy toilets can be quieted down by installing a new noncorroding plastic fluid-level control valve. It replaces the intake valve and controls water level by reacting to changes in water pressure. Directions for installing come with this inexpensive replacement.

Replacing a toilet

Toilet types

Washdown
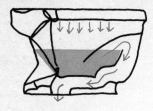

Reverse trap
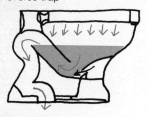

Siphon jet
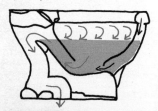

Toilets come in three basic types, according to how they flush. The *washdown* toilet (left, above) costs less, but it is noisier, uses more water and provides less trap-sealing protection. It usually has a rounded bowl. A *reverse-trap* toilet (left, center) costs more but flushes quietly and has a deeper trap seal and larger water area. Its elongated bowl shape is more comfortable. Reverse-trap toilets with front-of-the-bowl traps usually flush better. A *siphon-jet* toilet (left, below) is like a reverse trap but has a built-in orifice through which water jets to start flushing action rapidly. It uses the least water per flush and has the most positive flushing action. A siphon-vortex toilet is like a reverse trap with water entry holes set at an angle to make water swirl into and out of the bowl.

Toilets are also mounted in different ways, as shown below. Older styles have a wall-hung tank that feeds flush water through a chromed metal elbow into a floor-mounted bowl. On newer models, the tank mounts on a ledge at the rear of the bowl. In these popular close-coupled models, all water passages are internal and a gasket between tank and bowl prevents leaks during flushes.

Even more modern is the one-piece integral tank/bowl toilet. Since its tank is not much higher than its bowl, flushing action tends to be more sluggish but quieter than a close-coupled model's. A one-piece unit is also less likely to overflow, since the tank's overflow outlet is at a slightly lower level than the bowl's rim.

Quite modern is the wall-mounted toilet whose tank hides behind a removable access panel in the wall. Its bowl bolts to a metal hanger mounted to wall framing behind the toilet. It's easy to clean under since the bowl doesn't touch the floor. But to get at the tank, you have to remove the panel and work in the confines of the wall.

You can replace one type of toilet with another—a washdown model with a siphon-jet, for example. But avoid replacing a floor-mounted toilet with a wall-mounted one. It requires extensive replumbing.

Wall mounted

Close coupled

One-piece, integral model

Removing the old toilet

Replacing a toilet takes half a day if you have the new toilet on hand before you begin. Most toilets are built to what is called a 12-inch rough-in. This means that the toilet will mount over a waste pipe centered—as most are—a foot away from the wall behind the toilet. The center of the two porcelain caps on the base of the bowl (or of the rear pair, if there are four) should also be 12 inches from the wall. If it isn't, ask your supplier for a toilet that fits your dimensions. Plan to clean up or paint the wall behind the toilet, once the old toilet is removed. If the new bowl's base is smaller, you may also want to redo the floor where the old bowl was.

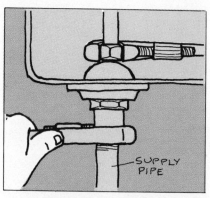

1. Shut off water supply to old toilet and hold down handle to flush it completely. Sponge out water remaining in the tank and bowl. Disconnect water supply pipe from the base of the intake valve.

2. Apply penetrating oil to hold-down bolts connecting the tank and bowl of a close-coupled unit; remove them and lift tank off the bowl. If tank is wall mounted, remove the elbow connecting it to the bowl and lift the tank from its wall bracket.

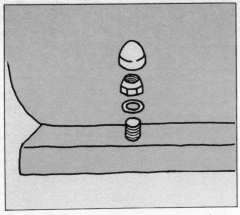

3. Unscrew or pry off the porcelain caps on the two or four nuts around the base of the bowl, chip off any plumber's putty from the nuts, and unscrew the exposed nuts.

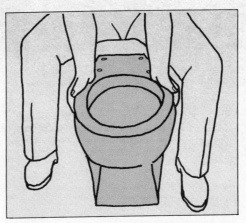

4. Straddle the bowl and rock, twist, and pull up on it until its seal with the floor flange breaks and you can lift it free. Keep it level while lifting so no trap water spills out. Clean old putty or wax out of the recess in the floor flange.

Need for a cover-up

Nothing attracts small tools like a hole in the floor. That's what I found when I put in a new toilet. Getting the old bowl up was no sweat but I didn't cover the hole in the floor until the new bowl could be put in place. Somehow one of my small wrenches "fell" into the hole. Lucky it was large enough to get stuck in the waste pipe where I could fish it out.

Fellow next door said to cover the hole with a small pie plate until you're ready to install the new bowl. Good idea. And, he added it might be smart not to leave tools in pockets they can fall out of, or on floors where they can be stumbled over or kicked into holes.

Practical Pete

Installing the new toilet

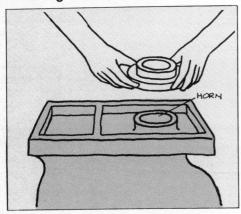

1. Upend the new toilet bowl on a padding of newspapers and slip a new wax toilet gasket, tapered side up, over the waste-water outlet, which is called the horn.

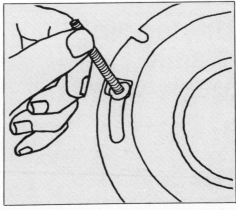

2. Slide a pair of new hold-down bolts into the slots in the floor flange, packing them with putty if needed to hold them upright. Lay a rim of plumber's putty around the bowl's flange. Pick up the new bowl, turn it right-side up, and lower it gently over the upright bolts in the floor flange. The wax gasket should meet the flange squarely.

Replacing toilet seats

The trick to replacing an old toilet seat is getting off the corroded bolts that hold the old one. Dose them overnight with penetrating oil. Fit a deep-throated socket wrench to the nuts under the bowl and try to turn counterclockwise. If the nuts won't budge, saw off the bolt heads with a hacksaw, being careful not to scratch the bowl.

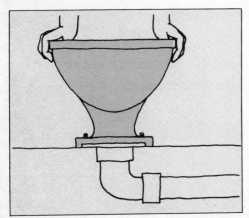

3. Twist and rock bowl gently as you press it down firmly. As excess wax squeezes out, the bowl will settle into position with its base against the floor. Install washers and nuts on the hold-down bolts and get them snug but not tight enough to crack the bowl. Cover the nuts with porcelain caps filled with plumber's putty.

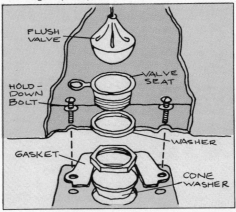

4. On a close-coupled model, position the big gasket around the bowl's flush-water inlet and set the tank on it. Install tank hold-down bolts and washers and snug them down. Connect water supply pipe to base of intake valve. Turn on water and test flush. If the new tank is wall mounted, install its bracket and hang it. Install elbow between tank and bowl and complete connections and testing.

Trace the bowl outline on stiff paper or cardboard so your supplier can provide the right shape replacement seat. Those with plastic hinges and mounts won't corrode. If you buy one with metal bolts, taping the bolts with Teflon tape before installing will make later removal easier.

TOILETS 35

5. PROBLEM SOLVING

Unclogging drains

If only one fixture is clogged, clearing its trap and waste connection should free the drain. If several fixtures are clogged, the blockage is in the branch drain serving them, or in the main house drain or its connections to the sewer. Unclogging these is explained on pages 38 and 39.

To clear a drain, you can use a force cup plunger or a drain auger (page 15), or pour in a chemical drain cleaner. Sometimes even a coat hanger or garden hose will do the trick. Using drain cleaners sounds easy, but some are so caustic that they should not touch skin, eyes, or clothing. Don't use them on a completely blocked drain. If they fail to clear it, you will have to handle and dispose of the caustic mixture when you clear the trap mechanically. But a once-a-month dose of drain cleaner in a cleared drain may keep it from clogging up as often as it otherwise would.

Whenever you clear a blockage with a plunger, auger, or other tool, flush the drain with hot water and detergent.

Freeing sink or lavatory drains

1. To clean out hair and grease caught in the stopper, lift out the stopper. Some must be rotated before they will lift out. If the stopper won't budge, it's held by a pivot rod and retaining nut under the sink. Put a pail under the sink drain. Then loosen the nut as above and pull the pivot rod back to release the stopper. If the stopper is clean and the drain is still blocked, try a force cup plunger, as in step 2.

2. Spread papers around the sink. Remove strainer or back pivot rod holding a pop-up stopper out of waste pipe. Close any overflow opening with wet rags. Leave enough water in the basin to cover the cup. Smear petroleum jelly around cup's rim and seat cup firmly over center of drain hole. Without lifting the cup, pump up and down with short, rhythmic strokes eight times, and then jerk it up off the bowl sharply. If blockage persists, repeat twice more before trying an auger.

3. Insert the boring head of a drain auger into the pipe and use thumbscrew to set the handle three or four inches from the drain opening. Rotate the handle slowly as you push the auger in. As the handle nears the opening, loosen it, move it back, and re-tighten. If you hook onto the clog, move auger back and forth slowly while continuing to rotate it in the same direction, then withdraw it. If this doesn't pull out the clog, crank the auger back into the clog until it breaks through. Then work it back and forth several times.

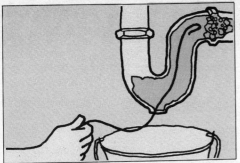
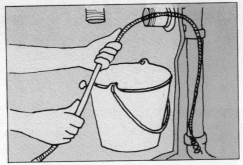

4. If the first go with the auger doesn't do the trick, put a pail under the trap's cleanout plug and unscrew the plug. Straighten a wire coat hanger and form a small hook on one end. Use this to probe up toward the basin drain hole and back toward the wall stub-out. If you can't hook onto the clog and draw it out, work the drain auger up into the openings as you did the coat hanger. If this doesn't do it, you'll have to remove the trap. You would have to do this anyway if the trap had no cleanout plug.

5. Loosen the trap slipnuts and take off the trap (page 26). Clean it out with rags wrapped around a stiff wire, a bottle brush, and water and detergent. Replace any that are worn out or corroded. If the blockage wasn't in the trap, run the drain auger into the open end of the drain pipe that goes into the wall or floor. If the auger moves in freely until it has gone far enough to meet the vertical soil stack, the blockage is probably in the main drain system (page 38).

New wrinkles

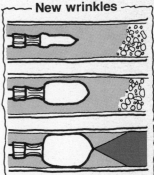

Plumbing suppliers now carry a low-cost rubber hose nozzle for clearing drains. Feed it into the drain and, when water is turned on, the nozzle expands to seal the drain more tightly than rags or a makeshift cover can do. Meanwhile a stream of water is directed toward the clog from the front of the nozzle, and the resulting pressure usually clears the clog.

Freeing tub drains

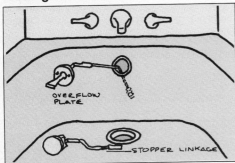
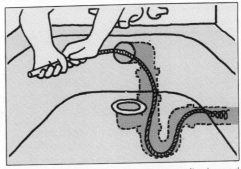

1. To clear blocked tub drains, you must first remove the strainer over the drain opening, and the pop-up or trip lever and stopper linkages from the drain and overflow pipes. Unscrew the strainer or lift out the metal stopper and its linkage. Unscrew the overflow plate and draw out the lift linkage of the trip lever or pop-up drain. Clean off any hair and other debris from the strainer, stopper, or linkage parts.

2. If the stopper or its parts weren't clogged enough to cause blockage, close the overflow hole with rags or masking tape, and try the force cup plunger. If that fails, feed an auger through the overflow opening rather than the tub drain. It has a better chance of working its way through the P-trap below the floor.

Stopgaps and copouts

Apartment dwellers faced with frequent sink backups and a landlord who won't listen can now install a small plastic check valve. It fits between the sink drain and its waste pipe, opening to let water drain out but closing to block off waste water backing up into the sink.

Freeing floor drains

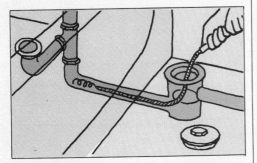

3. Some tubs in older houses have cylindrically shaped drum traps located at floor level beside the bathtubs. Their metal covers can be unscrewed and lifted off to expose a lower opening going to the tub, and a higher one going to the main drain. Feed an auger through the lower opening first, then back through the higher one. If the trap's gasket is worn, replace it while you have the cover off.

Force cups and augers can be tried on floor drains but a garden hose may be more effective if the blockage is beyond the trap. Remove the strainer and screw a garden hose to the nearest faucet. (Get a threaded adapter for a lavatory or sink faucet.) Feed the hose into the drain as far as it will go and pack rags tightly around it at the drain opening. Or, slit a sturdy plastic kitchen bowl cover and fit it around the hose to seal the drain opening (above). While you hold down the rags or plastic cover firmly, have someone turn the water full on and full off repeatedly. The short bursts of water should clear the clog.

Caution: Always remove and clean a hose used to clear drains. Leaving the hose in can result in hazardous back-siphonage (page 23).

PROBLEM SOLVING 37

Freeing toilet drains

Clogged toilets are cleared with force cup plungers or augers that are specially designed for the task. The plunger has an extended lower cone that more nearly fits into a toilet's built-in trap opening. The auger—called a closet auger—has a long tube with a curved tip that is placed close to the trap opening. Cranking and pushing the handle feeds auger wire from the tube into the opening. Pulling the handle back retracts the wire into the tube. Follow the steps below in using these tools.

1. The toilet bowl should contain just enough water to cover the plunger cup. Don rubber gloves and fit the cup over the large opening near the bottom of the bowl. Without lifting the cup from the bowl, pump up and down 8 or 9 times with quick, rhythmic strokes, then jerk the cup up sharply. If the bowl water disappears, you have probably freed the drain. Pour in some water as a test. If the drain is still clogged, plunge away a few more times before using the auger.

2. With the bowl half full of water, place the curved tip of the closet auger tube so it faces into the drain opening, push in and turn the handle to feed the auger's boring head into the drain. Use the same pushing and withdrawing motions used for clearing sink or lavatory clogs with a drain auger.

If the blockage is obviously in the built-in toilet trap and the auger won't clear it, you'll have to remove the bowl (pages 34-35) to get the clog out.

Freeing house drains

When a clog occurs in the main drain or its branches, every fixture drain above the clog will be affected. Turn on a faucet or flush a toilet and waste water may suddenly appear in the bathtub or shower, which have lower drains. Or a clothes washer in the basement may flood. You'll have to clear the clog from the branch or main drain.

If drain water is still flowing out—but sluggishly—try running water after a long time of not running it. If the drain backs up again quickly, the clog may be close enough to reach with an auger through the nearest fixture drain. If it takes a long time for the water to back up, the clog is probably far down in the waste system. Or, if you've noticed unpleasant odors in the house, the vent system that connects to the waste system may be blocked. Blocked vents also create a sluggish flow.

To clear the vent and the main soil stack to which it attaches, rent a 50-foot long electrically powered auger from a local tool rental firm or plumbing supplier. Feed this down through the stack vent on the roof, as at right. Alternately push and withdraw it if you meet resistance, until the auger breaks through the clog. When it does, flush the stack with a garden hose.

The rented auger should reach down to the main, Y-shaped cleanout from the roof. If you encountered no blockage, the clog is somewhere beyond the main cleanout plug. Try step 1, opposite.

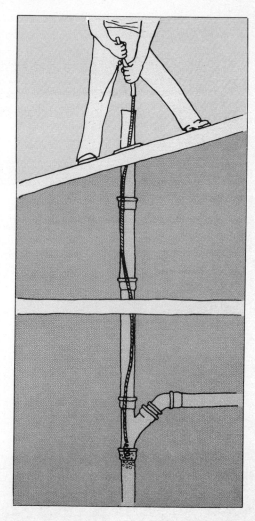

New wrinkles

Rather than buying a regular force cup plunger for sinks and lavatories, and a second one with the extended cone designed for toilets, you can now get one with a foldout flap that forms a cone when you need to plunge a toilet. One convertible force cup plunger for less than the price of two. Makes sense.

38 PROBLEM SOLVING

1. Spread several layers of newspapers around the main cleanout plug and station a pail under it. Loosen and remove the plug as above. But keep it handy to help slow down any flow of water that might come out. When the flow has subsided, go to step 2. (If you can't loosen a plug with a wrench, nick its edge with a chisel, hold the blade in the nick and drive the plug counterclockwise with a hammer.)

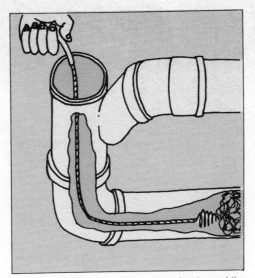

2. Feed an auger through the opening toward the main house trap, as above. If this clears the clog, flush line with a garden hose and recap the opening after coating the plug's threads with pipe dope. If you still have blockage, the clog is in the U- or Y-shaped main house trap with its adjacent cleanout plugs. Or it's between the main trap and the town sewer connection. Try step 3.

3. Spread layers of newspapers round the trap and slowly loosen the plug nearest the outside sewer line as above. If no water seeps out as you loosen, the clog is in the main house trap. If you see the clog, poke a small hole in it with a stiff wire and let water drain through gradually. When the flow subsides, open the adjacent plug and clean trap thoroughly with an auger, a stiff brush and a hosing.

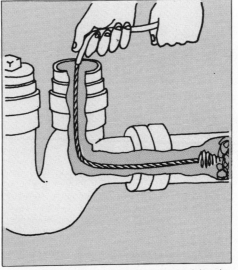

4. If water seeped out when you loosened the plug nearest the sewer, remove the plug and quickly feed an auger, or better yet, a sewer rod or tape into the drain leading to the sewer, as above. If this doesn't remove the blockage, recap the trap and consult with town officials, or with a professional plumber if you have a septic system.

Rescuing valuables

Valuables dropped into a drain may not be lost forever. Shut off all water immediately and make sure no one flushes a toilet until the prize is recovered. On a sink or lavatory, lift out stopper gingerly; the valuable may be caught in it. If not, shine a flashlight down the opening; if you see the valuable, you may be able to fish it out with needlenose pliers.

If it's not there, place a dishpan under the trap cleanout and remove the plug. Form a hook on an end of coat-hanger wire. Work this up toward the drain opening and pull it down gently. Then remove the trap and probe through it. For a toilet, wrap your arm in a plastic bag and work coat-hanger probe through top of trap and back toward you along the bottom of the trap to either hook or roll the object to you.

Tap horizontal drainpipes between the fixture and the main house trap with a soft-faced hammer or mallet while listening for a rattle inside. If the valuable is metal and the pipes are not, you may be able to rent or borrow a metal detector to help locate it.

As a final step, remove a main house trap-cleanout plug and cut a circular piece of wire screening whose mesh is small enough to catch the valuable. The screen diameter should be larger than the pipe's so it can be fitted tightly in the pipe. Turn on water and run either the sink faucets or the toilet the valuable fell in. The flow of water may move the valuable to your trap. Whether it does or not, remember to remove the wire screen and recap the drain.

Solving frequent clogging

A house sewer that clogs ever more frequently may be filling with tree roots. You can rent an electrically powered root auger that will clean them out, but if you do, get careful directions on its use and follow them. Misused, this cutting tool can chew right through some of the softer kinds of sewer pipe.

Throughly wash and dry any equipment used for drain cleaning before storing. Oiling will keep it from rusting.

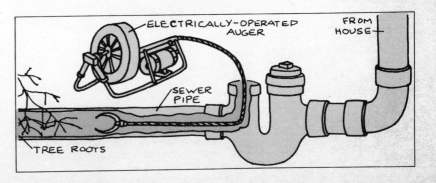

Fixing leaks

Tip: Sometimes it's hard to tell whether a toilet tank is leaking or just dripping condensation. Add some vegetable dye to the tank water, wait an hour, and touch the bolt tips and nuts under the tank and on the elbow connecting the tank to the bowl with white tissue. Where dye color shows, there's a leak. Shut off water, flush tank and mop out with a sponge. If a hold-down bolt leaked, tighten and re-test. If leak persists, remove bolt and replace washers. If leak is at base of intake valve, install new fluid-level control valve (page 33). Loosen leaking elbow slipnuts, replace washers, coat threads with silicone rubber sealant or self-forming packing and re-tighten.

If leak is around flush-valve seat (page 35), remove hold-down bolts and disconnect water supply and intake valve. Lift tank from bowl, lay it down and remove lock nut on valve seat. Replace round and cone-shaped washers and remount the tank.

Loose or poorly fitted joints and corroded pipes and fittings all cause leaks. Some joint leaks can be stopped by replacing washers, if any, and tightening or resoldering. On others, corroded fittings must be replaced. Where pipes or storage tanks leak from corrosion or mechanical damage, the leaks can sometimes be stopped with commercial patching devices. If pipe or tank has corroded from inside enough to leak, it's best to replace it.

Burst supply pipes are scary but shutting off the nearest stop valve will stop the flow. If it's a hot-water pipe, close the stop valve above the hot-water heater. If you can't locate a nearby stop valve, close the main shutoff valve (page 6).

Water dripping from or staining a ceiling usually means a leak in the water system above. Check all fixtures for visible leaks. If you find any, replace gaskets, washers or worn parts, retighten or resolder faulty connections and test for further leakage.

If no fixtures were leaking, the leak may be behind the wall. With the water on, listen upstairs along the wall where you know water pipes run for a hissing, gurgling or muffled dripping sound. It should be loudest when you're closest to it, with no signs of leakage above it. Some plumbing supply stores sell simple sound-amplifying devices that help pinpoint leaks.

Don't try to patch a pipe leaking behind the wall or one that has burst. Replace it.

1. Small, pinpoint leaks in pipes can be sealed with tightly wrapped plastic tape but fix small joint leaks with epoxy paste. Shut off water, drain pipes and clean and then dry the outside of the pipe with a hair dryer before applying the epoxy. Let it cure overnight before turning on the water.

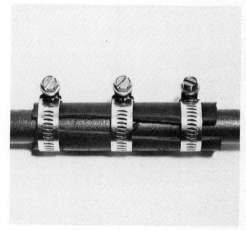

2. Larger cracks can be sealed temporarily by binding a section of hose tightly around the crack with twisted picture wire or, better yet, worm-drive hose clamps, as above. Use enough wire or clamps to keep pressure uniform all along the clamp.

3. As soon as you can, replace a homemade hose clamp with a rubber-lined pipe leak clamp or sleeve of a diameter that fits the damaged pipe. The sleeve can be cinched tight around the cracked area with screws and nuts.

4. Small leaks in tanks can be temporarily plugged with self-tapping repair plugs (see Table, opposite). These are screwed into the rusted-out hole until the plug's gasket seals the leak. But don't count on this fix lasting indefinitely.

40 PROBLEM SOLVING

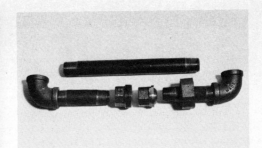

5. Leaks may be visible on exposed piping or they may show up as stains on walls or ceilings. If the sweep hand or one-cubic-foot dial pointer on your water meter moves when all fixtures are shut off, the leak is from your plumbing and not a hole in the roof or wall.

6. To replace a leaking section of galvanized threaded pipe, get two shorter sections of matching pipe and a union which, when put together, are the same length as the pipe being replaced. Hacksaw a section out of the old pipe and remove its pieces. Then fit the new pipes and union in position as shown. Remember to tape threads and tighten nut with two opposing wrenches.

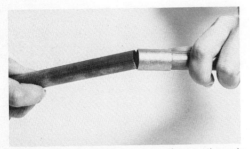

7. To repair a small leak in copper pipe, cut through pipe where leak is, coat the inside of a clean copper slip coupling with flux, and spread cut pipe ends enough to slip coupling onto one end of pipe. Clean and flux pipe ends, center slip coupling over cut, and sweat solder each end of the coupling to the pipe. Wrap wet rags around first coupling end soldered to keep it from melting as second end is soldered.

8. Leaks in cast-iron waste pipes can often be stopped by wrapping pipe in a rag soaked in silicone sealant. If a lead-caulked waste joint leaks, tamping it down with a hammer and chisel or sealing it, after drying, with epoxy paste should stop the leak.

Troubleshooting leaks

Problem	Solution	What it takes
Water spurts from pipe.	Install clamp-type patch until pipe can be replaced.	Patch and wrench or screwdriver.
Water leaks from joint.	Tighten threaded pipe joint. Drain and resolder sweat-copper joint. Cut out and replace solvent-welded plastic joint.	Pipe wrenches, hacksaw, pipe cutter, knife.
Water leaks from tank of hot-water heater.	Remove insulation jacketing and install repair plug, until heater can be replaced.	Tin snips, drill, bit, wrench.
Water drips from a ceiling or along foundation.	Locate hidden leak and remove wall or ceiling material to expose it for repair. Replace pipe or fitting.	Carpentry and pipe tools.
Continually wet spot in ground above service entrance pipe.	Have ground-key valve turned off, dig hole above leak and replace leaking pipe or fitting.	Shovel and pipe tools.
Water leaks at water heater's T&P relief valve pipe.	Check house pressure. If too high, install pressure-reducing valve. Valve off some water to clear sediment. If that doesn't stop leak, replace valve with one rated according to local code.	Plumbing tools, O-100 psi pressure gauge, pipe wrenches.
DWV system leaks at cleanout fitting.	Tighten fitting.	Wrench.
DWV system leaks at pipe or joint.	Make permanent pipe repair. Lead-and-oakum caulked joints may be repaired by repounding lead with caulking tool.	DWV piping tools, joint caulking.
Trap leaks at slipnut.	Tighten slipnut. Remove trap and install new O-rings. Or, as a temporary repair, coat with silicone rubber sealant and wrap with electrical tape.	Trap wrench, O-rings or silicone sealant and tape.

Curing condensation

Water condenses on cold metal pipes when humid air flowing over them is cooled below its dew point. If you don't want to solve the problem with a dehumidifier, you can keep the pipes from dripping by wrapping them with insulation. Several types are available: fiberglass, aluminum-and-vinyl and other plastic-tape wraps, or preformed plastic sheathing. Tapes are wrapped around pipes and fittings like a tight bandage. If they are not self-adhering, fasten the ends with string, tape or household cement. Preformed sheathing is slit on one side to snap around pipe and then tape tight. Both types eliminate condensation and provide some insulation. But some are meant primarily to stop condensation and should not be used on hot-water pipes. Ask your plumbing supplier to provide the type you need.

Torch song
Boy, did I get burned the first time I tried to thaw a frozen pipe. Fellow who knows the ropes told me I didn't know how lucky I was just to get burned—and not have a pipe burst in my face.

The right way, he said, is to use a flame spreader nozzle and never to let the pipe get too hot to touch with your bare hand. Always start near an open valve or faucet and work away from it until water begins to flow. This lets any steam pressure from heating escape harmlessly through the faucet. Otherwise it could form a pocket blocked by ice and burst the pipe.

Some other cautions: Always keep a sheet of asbestos between the flame and any nearby framing or walls; and it won't hurt to have a fire extinguisher handy just in case.
Practical Pete

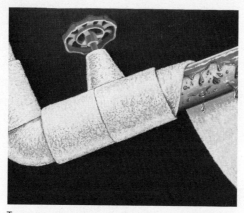

Tape wrap prevents condensation and dripping.

Slit sheathing snapped around pipe stops the drip.

Preventing freezing

The best cure for freezing pipes is prevention. A wrapping of electric heat tape should keep pipes from freezing. To make the heat tape work better and last longer, wind a plastic-tape wrap over it.

The water supply to hose bibbs and other outside piping should be cut off and the pipes drained in winter to prevent freeze-up. If a power failure and subfreezing weather catch you with uninsulated pipes, turn faucets on to a trickle and wrap pipes with several layers of newspapers tied with string. If the power failure and cold snap continue for some time, drain the house system and pour antifreeze into the fixture traps as explained on page 90.

Thawing pipes

When unusually cold weather or a power failure freezes up pipes you never expected to freeze, what to do? If you have power, placing an electric hair dryer on the pipe near an open faucet or drain valve, or wrapping an electric heating pad around it will thaw the pipe (see opposite page, above left). Or you can tie towels or rags around the frozen pipe and pour boiling water over them until the ice melts (opposite, above center). If you have the time and think the pipes may freeze again, buy electric tape and use that to thaw the pipes and keep them from refreezing.

A quicker method that requires caution but works even when power fails is to use a propane torch. Fit it with a flame spreader nozzle and play it along the pipe (opposite, above right). But be sure to observe all safety precautions (see Torch song at left).

Frozen pipes behind walls or ceiling can often be thawed by aiming a heat lamp at them. Keep it about 8 inches away to avoid marring or scorching the wall or ceiling.

For buried pipe that has frozen, call local arc welding shops and ask whether they have any experience in thawing pipes by sending electrical current through them. This is tricky work that should only be done by experienced operators.

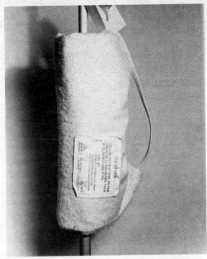
Use a heating pad near open faucet to thaw frozen pipe.

Pour boiling water over pipe wrapped with towels or rags.

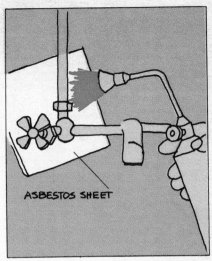
Torch with flame spreader nozzle.

Curing water hammer

When fast shutoff valves—like those on faucets or clothes washers—stop incoming water abruptly, it may make pipes vibrate noisily. This is known as water hammer. It is annoying and it can damage the water supply system.

The cure is to have air chambers (page 7) attached to the water supply lines of every fast shutoff fixture. Most newer homes have them. Adding them to behind-the-wall fixture piping in older homes is tedious and costly. Owners of older homes will probably wait to do it until some remodeling gives them an excuse to get behind the wall. In the meantime, installing the add-on chamber shown below will provide some noise protection in older homes.

Sometimes water hammer occurs when existing air chambers become waterlogged. The first step is to turn off all faucets slowly to stop the hammer. A longer lasting cure is to drain the water supply system and refill it. This replenishes the air cushions in the chambers. Shut off the main supply valve and open all water faucets. Draw air into the system through the faucets by opening the drain valves at the bottom of the water heater and the outdoor hose spigots. Then close all faucets and valves and refill the system. If water hammer has disappeared, its cause was waterlogging. Whenever it reappears, another draining will stop it.

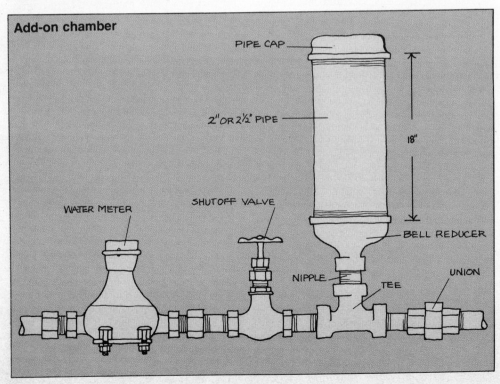
Add-on chamber

PROBLEM SOLVING 43

6. PIPING

Pipe fittings

Typical fittings used in water supply piping include: **1, 2,** and **3.** Sweat-solder copper tees; **4.** Standard copper coupling and, below it, copper slip coupling; **5.** Threaded steel union; **6.** Steel coupling; **7.** All-thread close nipple; **8.** CPVC tee; **9.** CPVC 45-degree elbow; **10.** CPVC cap; **11** and **12.** ¾- and ½-inch 45-degree copper elbows; **13.** Thread-to-sweat solder adapter; **14.** Threaded nipples; **15.** CPVC coupling; **16.** CPVC 90-degree street elbow; **17.** CPVC 90-degree elbow; **18.** CPVC adapter to threads; **19.** Left to right, flexible pipe tee, adapter and 90-degree elbows. **20.** Brass flare fittings: (a) elbow; (b) tapped elbow; (c) adapter; (d) tee; **21.** Brass hose adapter; **22.** Threaded brass 90-degree elbow and pipe section; **23.** CPVC pipe with brass adapter attached.

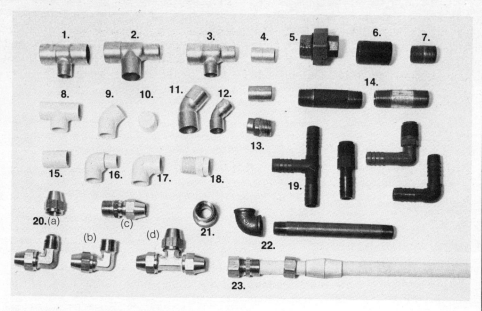

Vinyl water-supply piping

Joining CPVC solvent-welded pipes

CPVC plastic pipe may be cut with a hand saw, hacksaw, or a tubing cutter with special wheel.

Remove the burr from the inside and outside of pipe. Do the inside with a knife, the outside with sandpaper.

First apply cleaner to both mating surfaces; then brush solvent liberally onto the pipe end and fitting socket.

Immediately join the pipe and fitting with a slight twist and hold for a few seconds.

Hot/cold vinyl tubing is ideal for homeowner installation. It's low in cost, lightweight, easy to use, and it lasts and lasts without corrosion or scale buildup. It comes as rigid ½- and ¾-inch inside diameter tube in 10-foot lengths. Outside diameters are about ⅝ and ⅞ inch. You can cut CPVC tube with any saw and join it to a wide variety of CPVC fittings by easy, almost foolproof solvent-welding (see photos at left). Adapters permit connection to threaded and sweated pipes and fittings. CPVC water-supply tubing is rated to withstand 100 psi pressure at 180 degrees Fahrenheit. Do not confuse CPVC with PVC cold-water pipe. The latter is similarly rigid, but is not heat-resistant. Used for hot-water supply, PVC would soften and eventually fail.

If, for some reason, you must buy tube and fittings made by different manufacturers—and even if you don't—it's a good idea to test-check a fitting on the tube before solvent-welding. Solvent cement used to join CPVC tubes and fittings is different from PVC cement.

Perhaps the biggest drawback to CPVC tubing is its one-way joining process. Solvent-welding is so simple, that, unless you are careful, you can get into trouble. Once you join the tube and fitting, you're committed. A wrongly installed fitting must be sawed off and replaced.

Leaks are rare. If you do the job right, using lots of solvent on both tube and fitting, you need not worry about them. The chief cause of leaks is too little solvent. CPVC joints can be handled in minutes and are ready for water pressure in an hour.

Vinyl water-supply tubes expand and contract more with changes in temperature than do metal pipes. For this reason, all hot water adaptations to metal pipes and fittings should be made with what are called *transition fittings*.

A 10-foot length of CPVC hot-water tube will expand and contract about ½ inch with temperature changes. Be sure that the ends of pipe runs aren't installed tightly against framing. Leave the tube room to move. Likewise, the tube should be fastened to framing with its own special hangers that permit endwise movement without cutting or abrading. Use one support every 32 inches. Also, vertical lengths of CPVC tube taking off from mains and up through a floor or down through a ceiling need room to bend as the main moves with tempera-

Vinyl water-supply pipe sizes			
Nominal size	Outside dia.	Inside dia.	Wall thickness
½"	0.625"	0.489"	0.068"
¾"	0.875"	0.715"	0.080"

ture changes. Either slot out the holes for them or make sure they have at least 8 inches to flex before being bound into a hole in floor or ceiling.

Both plastic and threaded pipe fittings come in a type called *street*. Not the road in front of your house, street means simply pipe-sized. Fittings normally accept pipes in all sockets. On the other hand, a street fitting contains one socket that fits into another fitting.

CPVC water-supply tubing may be used to plumb an entire house or you can use it for extending a present system made of metal. In that case, the correct adapters will get the project started. At sinks, lavatories, and toilets, CPVC now can be adapted directly to riser tubes via fittings or fixture-shutoff valves. At bathtubs and showers, connections are made by adapting CPVC to the faucets by making use of transition fittings.

CPVC fittings

Special hangers for CPVC water-supply pipe are designed to hold the pipe to the framing, yet not bind or cut it. Be sure to use this kind of hanger.

Runs of CPVC pipe longer than 35 feet should be dog-legged by installing a 12-inch offset with two elbows. This allows for thermal expansion and contraction.

Caution: Vinyl solvent-welding cements contain volatile chemicals that should only be used in a ventilated room. Do not smoke while doping joints, and do not solvent-weld near an open flame. Also, don't spill solvent on plastic materials, such as some toilet seats. It dissolves the plastic.

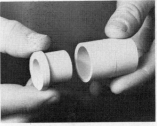

CPVC pipes can be reduced from one size to another by the use of bushings. The bushing is solvent-welded into the fitting; the smaller pipe solvent-welds into it.

New wrinkle

Polybutylene
Now you can get a flexible plastic tube for water-supply lines made of a new high-temperature thermoplastic polybutylene (PB). Use PB where soft-copper water-supply tubing is now used, such as for burial beneath concrete slabs where a single length averts having buried joints that might leak. PB also is fine for remodeling where pipes have to be snaked through walls, floors, and ceilings. Joints in PB tubing are made with adapters and CPVC or acetyl copolymer fittings: solvent-welding is not possible. Don't confuse PB pipe with polyethylene pipe. Polyethylene is flexible but can't carry hot water.

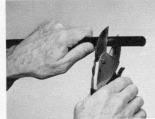  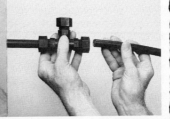

1. The tube is cut with ordinary garden shears. Tube cutter, saw with miter box, or even a sharp knife may also be used. 2. A metal insert or stiffener which comes with fittings is pushed into the tube end. 3. After the fitting nut is loosened by hand, the tube is pushed in and the fitting is tightened.

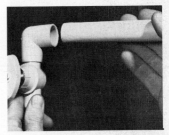

One type of CPVC line stop is called *street*, which takes either a ½-inch pipe in it or a ¾-inch fitting over it. It can thus serve as a reducing valve.

Comparison of plastic water-supply piping

Type of pipe	Type of use				Joins with	Form	Cost	Where to install
	Underground	Cold water	Hot water	Inside house				
Polyethylene (PE)	Yes	Yes	No	No	Slip-on clamped fittings	Flexible coils	Very low	Underground sprinkler systems, irrigation systems; generally for cold water, out-of-doors, and underground.
Polyvinyl chloride (PVC)	Yes	Yes	No	No	Solvent-welded fittings	Rigid lengths	Low	Same as for polyethylene.
Chlorinated polyvinyl chloride (CPVC)	Yes	Yes	Yes	Yes	Solvent-welded fittings	Rigid lengths	Moderate	For inside the house water-supply systems, cold or hot water. Ideal for exposed use in basements, utility rooms rooms where neat corners count.
Polybutylene (PB)	Yes	Yes	Yes	Yes	Mechanical O-ring fittings	Flexible coils	Moderate	For inside the house hot/cold water-supply systems. Best suited for hidden use, where appearance won't matter. For uses, see New wrinkle above.

PIPING 45

Threaded water-supply piping

Joining threaded pipes

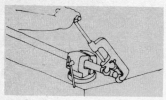

1. Cut steel and brass pipes with a pipe cutter, tightening a little each turn until the pipe severs. Remove resulting burr with a pipe reamer.

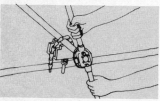

2. Use a pipe die inside a stock to cut threads. Start the die, guide first, and push and turn clockwise until the threads are established. Use lots of cutting oil on the pipe end and back up a quarter-turn every half-turn to clear the cuttings. Continue until one thread shows beyond the die.

3. To install a fitting, apply pipe dope or tape to the male threads and start the fitting by hand. When it starts getting tight, turn with a pipe wrench, the wrench's jaws faced to turn clockwise, and tighten.

Threaded galvanized-steel and brass water-supply pipes make the strongest, most damage-resistant systems which can be put in. Brass, while costly, gets around the corrosion and scale-buildup drawbacks of galvanized-steel pipe. Threaded pipes join to their fittings by tapered pipe threads. The farther the pipe screws into the fitting, the tighter the joint becomes. New pipes come threaded on both ends. In cut lengths of pipe, threads are made with a die—a thread-cutting tool—held in a stock.

Threaded pipes are sized nominally by their inside diameter. Because of their thick walls, outside diameters are much greater. The accompanying table shows how to judge threaded-pipe size by measuring a pipe's circumference. Popular house-water-supply pipe sizes are ¾ and ½ inch, with 1, ⅜, and ¼ inch used occasionally.

Pipe nipples of various lengths up to 12 inches long are widely sold. These are cut to length and come already threaded. Nipples short enough to leave almost no distance between fittings joined by them are called *close* nipples. So-called *short* nipples leave about ½ inch between fittings. Others come even-numbered, up to 12 inches.

Since most faucets and valves are already set up for threaded pipes, fewer adapters are required than with plastic or copper pipes. Galvanized pipes and fittings are

Neatest way to lubricate and seal pipe threads is with a wrapping of TFE white tape. Apply to male pipe threads before screwing them into a pipe fitting. Direction of wrap is important. Hold the pipe in your right hand and turn it away from you to pull the tape over the top. Give one full wrap plus some overlap and tear off. Use enough tension so the threads show through the TFE tape.

worked with toothed-jaw pipe wrenches. Cutting may be with a hacksaw, or more professionally, with a pipe cutter.

Each male thread should be coated with a good pipe dope or wrapped with TFE tape (see photo). When joining threaded pipes and fittings, it is possible to tighten them too much and crack the fitting. With new threads, tighten the joint until about three pipe threads still show beyond the fitting. Never tighten more than one turn after the last thread has disappeared into the fitting. You can gauge tightness by feel.

Judging galvanized pipe size			
Approx. circum.	Nominal inside dia.	Actual inside dia.	Outside dia.
1¹¹⁄₁₆″	¼″	0.36″	0.540″
2⅛″	⅜″	0.49″	0.675″
2⅝″	½″	0.62″	0.840″
3⁵⁄₁₆″	¾″	0.82″	1.050″
4⅛″	1″	1.04″	1.315″

Fixture requirements	
Fixture	Gpm flow
Bathtub	5–8
Tank toilet	2–3
Flush-valve	30–40
Laundry tub/Sink	5
Lavatory/Shower	5
Shower-water conserving	2
Garden hose	5–10

Comparison of Pipe Types

Pipe	Advantages and Disadvantages	Relative cost (CPVC/PB=1)	Sizes available
CPVC/PB	Very lightweight, smooth flow, nonconducting, resists corrosion and scaling, self-insulating, easy to use. Sensitive to over-temperatures, over-pressures, and hard blows. CPVC joints cannot be undone. Comes rigid or flexible. Cannot be used for electrical grounding.	1.0	½″ ¾″
Copper water tube	Lightweight, smooth flow, resists corrosion and scaling. Sensitive to over-pressures and hard blows. Offers rigid and flexible alternatives, choice of weights. Requires flame-soldering.	1.2*	⅜″ ½″ ¾″ 1″
Threaded galvanized steel, brass	Super strong, larger internal diameters for fast flow. Resists nails and most drilling, connects easily to threaded tub/shower faucets but requires threading. Brass is noncorroding, nonscaling, longest-lasting system. Requires more tools, skill.	1.7**	¼″ ⅜″ ½″ ¾″ 1″

*Type M tube. **Galvanized steel. Brass is many times as costly.

46 PIPING

Joining sweat-soldered copper tubing

Copper water-supply tubing is the easiest-working metal piping for house water-supply systems. Copper tube comes in three weights in nominal sizes according to inside diameter. Common house supply sizes are ½ and ¾ inch. Diameters of ⅜ and 1 inch are also used occasionally. No factory-made nipples are available. Instead, they are cut from tubing as required.

Type K tubing, the heaviest-walled, is too costly for most residential uses. Type L is the normal choice for buried service entrance lines and other below-ground purposes. Type M, thinnest-walled, is for aboveground house plumbing and for hydronic heating systems. Types K and L come in hard-temper straight lengths 20 feet long (sometimes sold cut to 10-foot lengths), and in 30- and 60-foot soft-temper coils. Type M tubing comes in hard-temper, straight lengths only. One difference between temper types is that while hard-temper tube cracks on freezing, soft-temper can take several freezings and thawings. Color-coding is: Type K—green; Type L—blue; and Type M—red.

Rigid tubes make a neater installation; flexible ones are best for underground uses and for unexposed remodeling work where they may be snaked into framing. Flexible tubes can be run long distances without fittings. Sweat-soldering may be used to join both hard- and soft-temper tubes by heating to about 400 degrees Fahrenheit and applying 50/50 lead-tin solder (see photos below). Soft-temper tube may also be joined by flaring (see page 14).

Some faucet manufacturers sell faucets that accept sweat-copper tubes directly without adapters. Otherwise, sweat-to-threads or other kinds of adapters are used to connect fixtures and appliances.

Copper tubing may be cut with a hacksaw or tube cutter. As with threaded pipe, the cutter does the best job. Be sure to thoroughly ream away the burr both inside and outside the tubing before making any joints. And keep in mind that the secret of successful sweat-soldering is cleanliness. Tube ends and fitting sockets must be scoured bright and shiny.

Sweat-soldered lines may be cast into mesh-reinforced concrete slabs or buried, though joint-free flexible tubing is preferred. Undergrounding of either type in cinders calls for first treating the tubing with asphalt to prevent corrosion. In any case, be sure to leak-check first. Always protect copper tubing from nail penetration by installing steel straps on framing.

New wrinkles

Instead of having to first apply flux to cleaned pipes to be joined and then coax solder to flow into the joint evenly, a new Swif Solder combines the flux with the solder. Just brush it on the cleaned parts to be joined as above, fit them together, heat the pipe adjacent to the joint until the solder flows (you'll see shiny, molten solder), and allow solder to cool. Move torch slowly to get an even spread of heat all around. When soldering unequal sizes, apply more heat to the heavier or thicker part. Remove any residue with a damp cloth or wet brush.

Water flow
(Type M copper tube)

Tube size	Gallons per min.
⅜"	2
½"	5½
¾"	12

Sizing of copper water-tube supply

Nominal size	Outside dia. Type K	Type L	Type M	Inside dia. Type K	Type L	Type M	Wall thickness Type K	Type L	Type M
⅜"	0.500"	0.500"	0.500"	0.402"	0.4030"	0.450"	0.049"	0.035"	0.025"
½"	0.625"	0.625"	0.625"	0.527"	0.545"	0.569"	0.049"	0.040"	0.028"
¾"	0.875"	0.875"	0.875"	0.745"	0.785"	0.811"	0.049"	0.045"	0.032"

Joining sweat-soldered pipes

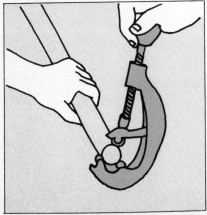

Copper tubing is cut with a tubing cutter. Tighten after each turn until the pipe has been cut through. Remove burr with a reamer afterwards.

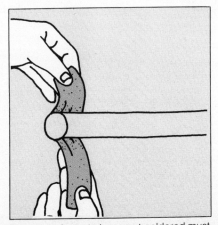

Copper surfaces to be sweat-soldered must be bright and shiny. Use fine sandpaper, emery cloth, or steel wool. Clean both the pipe and fitting socket, then apply a good, nonacid soldering flux to pipe end and insides of fitting.

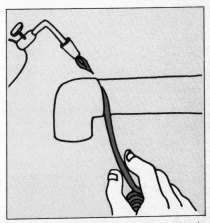

Join parts and heat with a propane torch. When solder will flow, remove flame and feed 50/50 wire solder into the joint until a bright fillet forms all around.

Running water-supply pipes

Probably the hardest thing about running pipes is remembering to allow for makeup. This is the distance taken up by a fitting minus the distance the pipe screws or slips into the coupling, tee, or whatever. Suppose you are running a supply pipe from a main line along the joists, and that it must come out of the wall centered 16 inches above the floor. You could take measurements and calculate exactly how long the pipe should be cut using the makeup table below. Subtract for distance taken up by fittings and add for socket makeup.

An easier way is to position the actual fittings at each end of the pipe where they will connect and measure what is called face-to-face between fittings. Then simply add for the socket makeup at each end—distance X for that pipe size times two. For example, if the face-to-face measurement is 23½ inches, then ½-inch threaded water-supply pipe joining the two ½-inch fittings would need to be 24½ inches long (23½ + ½ + ½). Threaded up, the fittings should position perfectly. But write down all the pluses and minuses to keep them straight in your mind.

Or, you can sketch out the pipe runs on paper, showing the measured distances the piping must traverse. Then, using the table and a pocket calculator, below, figure how long each pipe must be. The first time through, you'll make errors and have to recut some pipes. Don't worry. Order enough extra pipe to permit a few slip-ups.

The balance of water-supply piping work is merely a matter of drilling or sawing access for pipes and installing them with the proper fittings. Proceed in the direction of water flow, especially when working with threaded pipes where one end has to be free for tightening.

Drilling and notching

It's better to drill holes for pipes than cut out notches in joists or studs. Joists may be drilled anywhere along the span but the hole diameter should be no more than one-fourth the joist depth. Keep holes as small as possible and at least 2 inches away from edges. Joists should be notched in the end quarters only, never in the middle half. Notch should not exceed one-fourth the joist's depth. Studs may be notched up to 2¼-inches round or square, if strap steel is nailed over the notch opening to reinforce it. Notches up to 1¼ inch don't need the straps. Stud drilling rules are the same as those for notching.

Take great care to avoid boring into pipes or electrical cables hidden within the walls. Don't stand in water. When standing on concrete, lay down a block of wood to help keep you from being grounded. Use either a double-insulated electric drill or one that is grounded with a three-pronged grounding plug and grouding extension cord. Never defeat an electric drill's safety features by adapting a three-pronged grounding plug to a non-grounding outlet. Don't take chances with electricity.

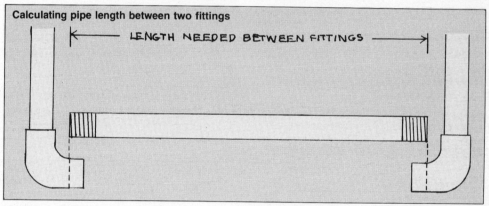

Calculating pipe length between two fittings

Makeup table for threaded fittings

Pipe size	Add distance — Pipe screws into fitting	Subtract distances				
		A	B	C	D	E
½"	½"	1⅛"	⅞"	1⅝"	1⁵⁄₆"	1¼"
¾"	½"	1⁵⁄₁₆"	1"	1⅞"	1½"	1⁷⁄₁₆"
1"	⁹⁄₁₆"	1½"	1⅛"	2⅛"	1¹¹⁄₁₆"	1¹¹⁄₁₆"

Makeup table for Genova CPVC fittings

Pipe Size	Pipe slips into socket	A	B	C	D	F
½"	½"	1⁵⁄₁₆"	11⁄16"	1"	1³⁄₁₆"	—
¾"	¾"	1¼"	1"	1⁵⁄₁₆"	1⁹⁄₁₆"	3"

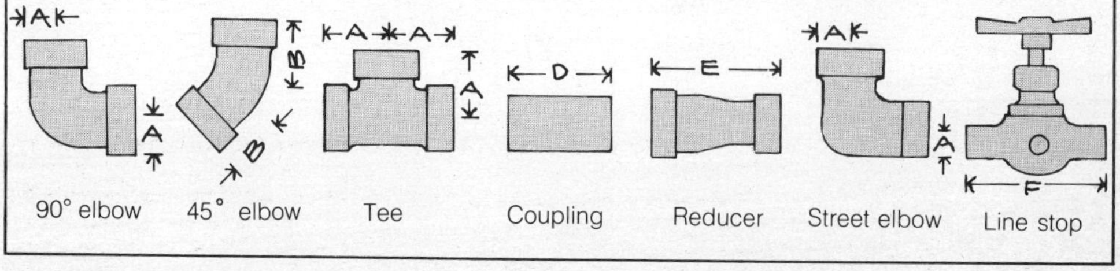

90° elbow · 45° elbow · Tee · Coupling · Reducer · Street elbow · Line stop

Typical DWV fittings

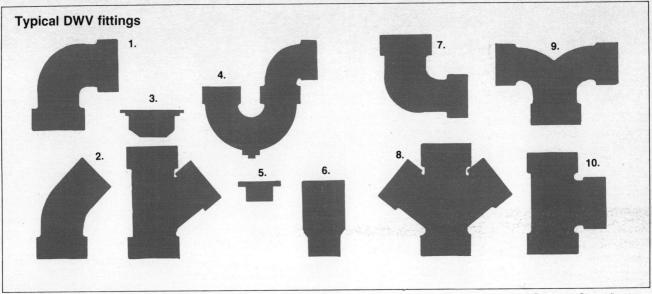

1. 90-degree sanitary elbow (hub and hub); **2.** 45-degree street elbow (hub and spigot); **3.** Cleanout wye with plug; **4.** P-trap; **5.** Closet flange; **6.** Reducing coupling; **7.** Closet bend (hub and hub); **8.** Double wye (all hub); **9.** Reducing double elbow; **10.** Sanitary tee.

Cast-iron DWV piping

Ask any plumber what is the best drain-waste-vent plumbing system, and he will likely say cast iron. Once installed, it cannot be surpassed. And installation is not difficult, except for cutting pipes to length, which is plain hard work unless you rent a lever-type cutting tool. The joining labor once connected with cast iron pipe disappeared with the introduction of the No-Hub system. It comprises simple hubless fittings and plain hubless pipes. Joining of pipes and fittings is by neoprene rubber gaskets (sleeves) and ribbed stainless steel shields held by worm-driven clamps. Pipes fit together so closely that no allowance need be made for makeup.

Another thing, a No-Hub joint, once made up, may be taken apart again to correct an error. A drawback, house walls containing 3-inch or larger No-Hub piping, such as behind toilets, must be furred out or built thicker than the standard 3½ inches to accommodate any fittings. However, this also is true of most other DWV fittings.

Hubless cast iron pipe comes in 5- and 10-foot lengths and 2-, 3-, and 4-inch inside diameters. No-Hub may be used vertically or horizontally, above or below ground. It is supported with a special clamp every 12 feet on vertical runs; every 4 feet on horizontal runs. Branch drains smaller than 2 inches are made with galvanized threaded steel, using galvanized drainage-type fittings. These are joined to 2-inch No-Hub via adapters. Traps and adapters are made in the 1½- and 1¼-inch galvanized pipe size. Pipe is standard; you can rent 1¼- and 1½-inch threading equipment.

No-Hub is particularly useful for connecting new DWV lines to older lead-and-oakum-caulked cast iron piping systems. Simply cut out enough of the old pipe to fit in a No-Hub wye, installing it with two No-Hub joints. Avoid caulking cast iron.

Joining No-Hub DWV

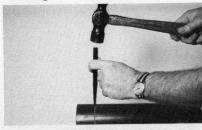 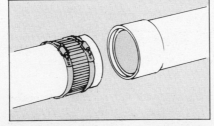 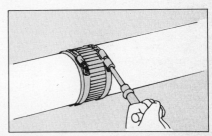

1. Groove pipe ⅛ inch deep all the way around with a hacksaw. Then, wearing safety goggles, work around the groove with a chisel until the pipe parts on the line. Hub-type cast-iron pipe is cut the same way.

2. To join a No-Hub pipe and fittings, put the parts together with a neoprene gasket on one piece and a stainless steel sleeve on the other. Slide the gasket over and center it on the joint. Then, slide the stainless steel sleeve over the gasket.

3. Finally, tighten the clamps alternately, bringing each a little tighter than the previous one until you reach a torque of 60 inch pounds (about what you'd tighten a small wood screw).

Plastic DWV piping

Plastic drain-waste-vent pipe is made in two materials: polyvinyl chloride (PVC) and acrylonitrile-butadiene-styrene (aren't you glad it's called ABS?). PVC is usually beige-colored, while ABS pipe and its fittings are black. Both kinds are joined by solvent-welding to fittings with solvent cement designed for that material. Both can be worked with ordinary tools like CPVC water-supply pipe. The chief difference in handling DWV plastic piping is its larger sizes. These call for use of a wider dauber or brush for joining.

Both ABS and PVC are safe for all the usual household wastes. They're also resistant to fungi, bacterial action, and adverse soil conditions. The simplicity and ease of installation of a plastic drain-waste-vent system make it the perfect choice for do-it-yourself plumbers.

Plastic DWV pipes should be supported every 48 inches—from every third joist. Stacks can rest on a wooden block at the bottom, as well as on the horizontal pipes that enter along their length. Pipes and fittings should be protected from subsequent damage by installing steel straps in locations where nails could penetrate. Plastic DWV pipe may be used above or below ground.

Ordinary roof flashings have trouble sealing around plastic DWV stacks. Thermal pipe movements tend to crack any sealant buttered around the flashing. Specially designed flashings that grip tightly to the vent pipe are best. These need no sealant to provide raintightness (photo, opposite).

Both PVC and ABS pipes and fittings come in 1½-, 2-, 3-, and 4-inch diameters and 10-foot lengths. Fittings for use with them should be labeled "DWV," signifying that they're designed for drain-waste-vent use. Schedule 40 fittings in the 3-inch size need 4½ inches of wall clearance to install. Lighter, smaller Schedule 30 3-inch DWV pipe and fittings will fit inside an ordinary 2x4-inch stud wall. Fittings for it mate with 2- and 1½-inch Schedule 40 pipes. Thus no adapters are needed between 3-inch Schedule 30 and smaller sizes of Schedule 40 pipe. The in-the-wall size of Schedule 30 is a great advantage in add-on plumbing, where the walls are already framed.

To ensure that fittings and pipes share the correct tolerances to fit together, do not use pipes by one manufacturer and fittings by another. And always test-fit the fitting on the pipe.

To get square pipe ends, saw pipes in a miter box. Before joining, examine pipe ends and fittings for deep scratches, abrasions, and hairline cracks caused by heavy impacts. If need be, cut off a damaged pipe end and discard it. Before applying solvent cement, chamfer the pipe end all around at a 45-degree angle with a file or sandpaper to help it enter the fitting cleanly. Don't use solvents around flames or cigarettes while you solvent-weld. Also, have the room well ventilated and keep solvent containers closed when not in use.

To tighten PVC or ABS threaded fittings, use a strap wrench. Pipe wrenches chew up plastic fittings. Use TFE tape or non-hardening pipe dope on the male threads before assembling. Don't put silicone rubber onto ABS piping.

For caulking shower drains around plastic DWV pipe, use special plastic lead seal compound instead of molten lead. If you must use hot lead, get a special high-temp adapter intended for a lead-caulked joint.

Stopgaps and copouts

Instead of sawing into a stack and installing a tee or wye to tap into it, you can get a solvent-welding saddle tee that lets you achieve the effect more easily. Simply weld the saddle tee onto the vent pipe in the desired location. After an hour, drill out the pipe inside the tee. A 1½-inch run takes off by solvent-welding into the saddle tee. (See below.)

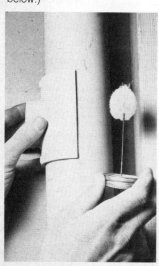

Schedule 30, 3-inch PVC pipe (right) will fit in a 2x4 stud wall. Schedule 40 PVC pipe (left) needs 4½-inches of wall clearance. See Wall thickness chart, page 54.

50 PIPING

Solvent welding

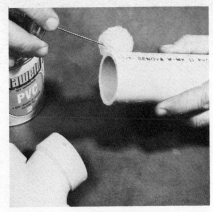 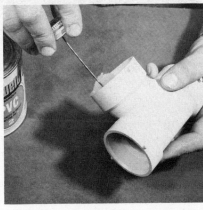 

1. After cutting, deburring, and test-fitting the pipe and fitting, same as for a CPVC water-supply pipe (see page 44), apply the proper solvent cement to the pipe end. Brush width should be half the diameter of the pipe.

2. Apply solvent to the fitting socket, also. Be generous and don't miss any spots. Be quick, too.

3. Without waiting, join the fitting and pipe full depth with a slight twist and align fitting. Hold for a few seconds. Wipe up any excess solvent cement with a cloth.

Running a vent stack

 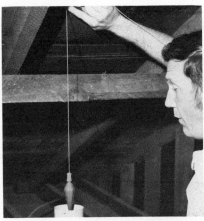 

1. Drill holes in the floor and ceiling plumb over one another to pass the vent stack piping. Large holes can be made by drilling a circle of smaller ones, then sawing out the remaining block of wood.

2. As you raise the stack through the attic, it's easy to clear framing members by off-setting it slightly, using a pair of 45-degree elbows. Use plumb line to fix vent location from offset.

3. A flashing at the roof keeps rain water from entering the house between the vent stack and roofing. This one is two-piece plastic. It snaps together, holding tightly around plastic DWV vent-stack piping.

New wrinkles

A new product you'll appreciate is the special waste-and-vent fitting. It takes the place of a sanitary tee under the floor behind the toilet. A pair of reduced side tappings allow 2- or 1½-inch waste pipes from lavatory, tub, or shower to drain and be vented at the same time. If you don't need the side tappings, solvent-welding plugs are available. The special waste-and-vent fitting also comes with caps that accept the vent stack at full bore. One also has an extra 1½-inch side tapping for another drain or vent. The fitting is available in single- or double-configuration for use with one toilet or two toilets back-to-back. It simplifies your early steps in building a DWV system.

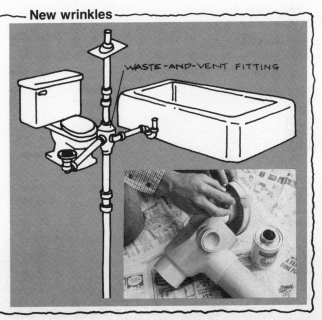

PVC or ABS?

Your local code and what is available may determine whether you use light-colored PVC pipe or black ABS pipe for your drain-waste-vent system. The two materials may even be joined to each other if you use ABS solvent. PVC solvent softens ABS pipe excessively, because ABS has less solvent-resistance. For joining PVC pipes and fittings, use PVC cement; for ABS pipes and fittings, use ABS cement.

Sweat-soldered copper DWV

Copper drainage tubing comes in drain-waste-vent sizes. It weighs about one-fourth what cast iron does. Tube/fitting sizes are: 1¼-, 1½-, 2-, 3-, and 4-inch diameters. Available length is usually 20 feet, though many dealers offer less than full lengths, if desired.

Copper drainage tube is suitable for above or below-ground plumbing. To prevent in-ground damage, however, it is better to use Type L copper tube in below-ground runs, since drainage tube has thinner walls. Otherwise copper drainage tube is like smaller-sized hard-temper copper water-supply tube (see page 47). Its outside diameter is always ⅛ inch larger than its nominal size. Fittings, though, are different, being drainage-type rather than water-supply fittings. Either copper or cast-bronze DWV fittings, which are made to mate with copper tubing, may be used. Copper drainage tube, which is color-coded yellow, is designed for nonpressure DWV application. But it may be used for pressurized piping if pressures are controlled. With soldered joints, about 55 psi is the maximum allowable safe working pressure.

Copper drainage tubing may be cut with a pipe cutter or with a hacksaw and miter box. Deburr the tube both inside and outside with a pipe/tube deburrer. Then test-fit the fitting onto the tube. Best clearance for proper soldering is a uniform 0.002" to 0.005" space.

Prepare a copper joint, as always, by scrubbing it shiny with steel wool. To facilitate solder entry, go an inch beyond where the tube will be covered by the fitting. And don't neglect to clean up the inside of the joint, too. Joint preparation should not be done with a coarse sandpaper or other harsh abrasive.

Copper drainage-tube joints are sweat-soldered much the same way as copper water-supply tubing (see page 47). But the extra-large sizes usually call for a larger propane torch, or for two torches. A pair of torches at opposite sides of the joint often works best. Heat up the entire joint before applying any solder, then fill it up all at the same time. Retorch the joint while solder is being applied, but always keep the torch or torches moving to maintain a constant temperature. Overheating will burn the solder. Less expensive 40/60 tin/lead solder will work for sweat-soldering. But 50/50 wire solder is recommended because it is easier to work with.

Don't forget the flux when sweat-soldering. Its purpose is to remove traces of oxides left after brightening, and to protect the surfaces from additional oxidation while they are being heated to soldering temperature. Fluxes are not meant to take the place of cleaning. So avoid types labeled "self-cleaning." Apply a light coat of a good paste flux to both surfaces to be joined. Since the best fluxes are mildly corrosive, don't apply with your fingers. Use a rag or small brush.

Fasten copper-drainage piping to joists at 40-foot intervals using perforated metal "plumber's tape." Vertical vent runs can be supported by the horizontal drain-and-vent pipes entering them. Copper flashings may be soldered to vent stacks at the roof. Generous application of roofing cement between flashing and vent pipes works, too.

How to avoid reheating a fitting

If you prepare all the pipes leading into a single fitting, you can sweat-solder them in one heating. This avoids having to heat the fitting again, which risks losing the solder in a part of the fitting that has already been completed. Be careful that you don't overheat the fitting. This burns the flux and oxidizes the copper so it will no longer accept solder. Do not clean too many fittings at one time, for they will oxidize if left more than three hours without soldering.

Joining copper DWV

1. As with copper water supply tubing, start sweat soldering DWV piping by cleaning mating surfaces with emery cloth or 00 steel wool. Clean 1 inch more of pipe than will fit into fitting. Wipe off dry and apply even coat of flux to all mating areas. Join pieces with a slight twist to spread flux evenly.

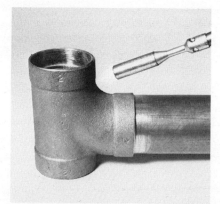

2. Apply flame all around first the heavier fitting and then the pipe. Heat both until tip of 50/50 wire solder touched to heated joint starts to flow. On large 3-inch DWV joints you may need big flame tip or two torches to heat joint adequately.

3. Melting solder is drawn into joint by capillary action. Make sure it flows into the joint all around to provide an even seal. Don't apply flame to solder. When joint is evenly filled, smooth solder and wipe off any excess with soft cloth before it cools completely.

Running DWV pipes

Here is the installation sequence for DWV: (1) underground drains and sewer; (2) building drain (from toilet tee); (3) main stack, secondary stack(s); (4) branch drains; and (5) fixture-waste and revent runs. Keep drainage runs with as few turns as possible. Long-turn elbows are preferable to short-turn ones, and two 45-degree elbows beat one 90-degree elbow. Wherever possible, use wyes instead of tees. Never change from vertical to horizontal flow with a tee. Use a wye instead. Be sure to include a cleanout, in an accessible location, for each horizontal drain run that cannot be augered through a nearby trap opening. This includes the house sewer.

Here are some rules for the installation of a DWV system: Always use drainage-type fittings for waste runs. Vent runs may be made with regular fittings or else with more expensive drainage fittings installed upside-down. In freezing climates, use vent-increasers on 1½- and 2-inch roof vents to protect against ice closure. Fixture waste pipes should enter a stack above a toilet. Entry in the same fitting as a toilet is okay, too. A fixture that is revented may be drained into the 90-degree elbow below the toilet (if it has a tapping).

Fixtures whose waste pipes enter below a toilet drain must be revented. Revent runs go directly out the roof or are connected into another vent stack at least 6 inches above the flood rim of the highest fixture emptying into that stack. A lavatory, which often drains into the waste line below the toilet, is normally revented into the main stack above the toilet.

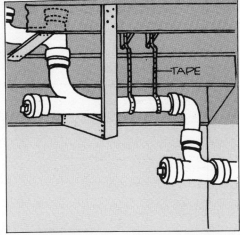

Drain-waste-vent piping should be well supported by the framing. Horizontal runs are hung from joists as shown, with supports 48 inches apart. Use perforated metal plumber's "tape" and wood bracing. Vertical vent runs can rest on wood blocks installed at the lower end.

A good place to begin DWV plumbing is at the floor beneath the toilet. When a sanitary tee—in this case, a special waste-and-vent fitting—is held in position, DWV piping may take off from it in three directions: toward the toilet, up and out the roof for the vent stack, and toward the building drain.

Haste makes waste

I learned that it pays to dry-assemble your plastic DWV system. It can save lots of headaches if things don't fit up the way you planned. This lets you make corrections and get everything right before welding the parts. Wherever a fitting-up problem is anticipated, dry-assembly is a good idea.

Practical Pete

Tapping into a drain run

1. Using shoulderless slip-couplings, tap into an old plastic run. Cut out the old pipe to make room for the made-up wye assembly with pipe extensions.

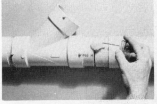

2. With the wye assembly held in place of the cut-out pipe, all four pipe ends are liberally buttered with solvent cement.

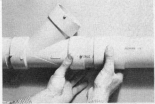

3. Then quickly slide the slip couplings over the joints until centered on them. Align the wye.

Comparison of DWV pipe types

Pipe	Advantages and disadvantages	Relative cost (Plastics =1)	Sizes available
PVC/ABS plastic	Very lightweight, smooth bore for full flow, resistant to corroding, easiest to install. Solvent-welded joints cannot be parted. Won't conduct electricity and may not be used for electrical grounding. Few tools needed. Resists all household wastes but affected by a few potent, flammable solvents. Needs protection from nailing. Easy to add onto. Widest selection of fittings. Use above or below ground.	1.0	1½" 2" 3" 4"
No-Hub cast iron	Strongest, quietest, most indestructible system available. No makeup allowances to bother with. Requires threaded galvanized piping for sizes smaller than 2 inches. Joints may be taken apart at will and replaced as often as desired. Easy to add onto. Use above or below ground. Difficult to buy, except from a plumber. Heavy to work with. Easily the best system, but also toughest to install.	2.3	2" 3" 4"
Copper drainage tube	Lightweight, smooth insides for good flow, resistant to corroding, not difficult to use. Is dented by hard blows. Not best for underground. Flame-soldering can be risky. Not melted except by intense heat. Few tools needed. The 3-inch size and smaller fit in a standard 2x4 stud wall. Needs protection from nailing.	3.2	1¼" 1½" 2" 3" 4"

Sizing DWV runs

In most single-family houses, drain-waste-vent runs may be sized according to the Rule-of-Thumb table below. But if two bathrooms are plumbed together, or a bathroom, kitchen, and laundry all use the same drainage pipe, then the flow produced by all the fixtures should be checked using the Fixture Units table. Then use the Pipe Capacity table to figure how large the drain piping must be to carry the calculated flow.

A fixture unit is a National Plumbing Code term used as the basis for figuring DWV pipe sizes. It represents a waste flow of one cubic foot a minute (about 7½ gal.).

After checking on drain pipe diameter, size the main stack. If only one or two toilets will empty into its base, a 3-inch stack is plenty large. Add up the fixture units for all house fixtures to see how large the main building drain must be to handle the load. Again, in most single-family residential uses, a horizontal 3-inch drain will do. Pipe sizes given in the Rule-of-Thumb table are minimums.

Fixture vent pipes should be the same size as their drain pipes. An exception is toilets, for which some codes permit 2-inch venting if it serves no other fixtures. Check your code. This important provision can get you around the need for thickening a wall behind the toilet to handle a 3-inch stack. However, if a 2-inch toilet vent is used, be sure to vent the other bathroom fixtures into another stack, perhaps a 1½-inch secondary stack installed just for them.

DWV pipe sizes—Rule-of-thumb (check local code)

Purpose	Pipe size	Purpose	Pipe size
Main stack	3" or 4"	Bathtub waste	1½"
Secondary stack	1½" or 2"	Shower waste	1½" or 2"
House drain	Matches stack	Tub/shower vent	1½"
House sewer	3" or 4"	Sink waste/vent	1½"
Toilet waste	Matches stack	Laundry waste	1½"
Toilet vent-only	2"	Laundry vent	1½"
Lavatory waste	1½"	Basement drain	2"
Lavatory vent	1½"	Other drains	2"–4"

Pipe capacity

Pipe size	Fixture units permissible	
	Horizontal pipe	Vertical pipe
1½"	3	8
2"	6*	16
3"	20**	30***
4"	160	240

*waste only
**not more than two toilets on horizontal line
***not more than six toilets on one stack

Importance of pipe-sizing

Oversized drain pipes become "lazy," not being sufficiently charged with water to carry wastes away. So wastes accumulate until the pipe clogs. Drain pipes that are too small are easily clogged by waste build-ups, too. Size pipes correctly—neither too small nor too large—according to the criteria at the right.

Nominal sizes for fixture drains and vents

Fixture	Pipe diameter
Toilet	3 inches
Shower	2 inches
Bathtub, sink, lavatory, laundry tub, clothes washer, dishwasher, garbage disposal	1½ inches

What's horizontal? What's vertical?

The terms horizontal and vertical have different meanings to a DWV plumber than to a carpenter. No DWV pipe is strictly horizontal. Drainage lines should always pitch downward about ¼ inch per foot in order to drain. And vent lines that are "horizontal" run upward toward the vent stack at the same ¼-inch slope. So neither is strictly horizontal, even though they are called that.

Vertical vent or drain-vent pipes—stacks—are often truly vertical. But they needn't be. Offsets may be placed in them to clear obstructions, such as roof rafters. And there's nothing wrong with installing them slightly off vertical if necessary to accomplish some other purpose. Vertical simply means that they run in a general up-and-down direction.

Fixture units

Fixture	Fixture units
Toilet	4
Bathtub/shower	2
Shower only	2
Sink	2
Lavatory	1
Laundry tub	2
Floor drain	1
One 3-piece bathroom group (toilet, bathtub, lavatory)	6

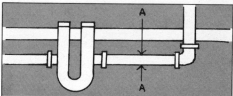

How large does the horizontal drain pipe at A-A need to be? First, since it handles wastes for a toilet, it must be at least 3 inches in diameter. Is that enough to handle everything else, too? To find out, calculate how many fixture units it handles. From the Fixture Units table, you'll find that the toilet gives 4, the bathtub/shower gives 2, and the lavatory gives 1 for a total of 7 fixture units. However, the table also shows that a single bathroom group actually contributes only 6 fixture units, so the actual total is 6. Next look in the Pipe Capacity table to see whether the required 3-inch pipe will carry 6 fixture units when running horizontally. The table shows it capable of up to 20 fixture units. Thus, 3-inch pipe is easily able to handle the load.

Wall thickness to accommodate vertical DWV pipes and fittings

Pipe size	Plastic Sched. 30	Plastic Sched. 40	Copper	Cast iron	Galv. steel
1½"	—	3"	2"	—	3"
2"	—	3½"	3"	4"	3½"
3"	3½"	4½"	3½"	5¼"	—
4"	—	5½"	4½"	6¼"	—

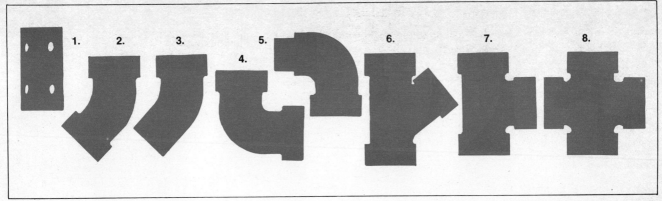

Typical sewer piping includes: **1.** Perforated pipe; **2.** 45-degree elbow (hub and hub); **3.** 45-degree elbow (hub and spigot); **4.** 90-degree elbow (hub and hub); **5.** 90-degree street elbow (hub and spigot); **6.** Wye (all hub); **7.** Sanitary tee (all hub); **8.** Cross (all hub).

Sewer pipe is different

Solid-walled sewer pipe is used for sewer lines between the house and the town sewer or septic tank. Loose-jointed or perforated seepage pipe is used in seepage fields and to dispose of roof runoff. It lets effluent, or runoff, percolate into the ground.

Plastic sewer pipe comes in two weights: heavy duty, and a lighter duty called "sewer and drain" pipe. Both are considered "soft" pipes that can be damaged by loads from above. Pitch-fiber sewer pipe is soft, too. Soft sewer pipes should be buried deeply enough under driveways so that car or truck weights won't affect them. Put them in a trench with a uniformly shaped bottom of unexcavated earth that is free of lumps or rocks that might cut or pinch the pipes. Dig out underneath couplings so that both pipe and couplings are supported.

Hard pipes—cast iron and clay—may go under driveways if there's enough earth over them to prevent crushing or bending from above. Cast-iron pipe may be suspended on concrete blocks placed in a trench beneath pipe joints. Uniform ground support is better, though.

A 4-inch sewer should handle all wastes from a single-family residence. A 3-inch line might do if it satisfies the total fixture flow estimated from table, opposite.

Joining sewer pipes

PVC, ABS, and rubber-styrene sewer pipes are joined by solvent-welding. The process is the same as that for plastic DWV piping. Use perforated pipes for septic seepage lines.

Pitch-fiber pipes are joined by tapered ends that fit tapered couplings. Tapping with a heavy hammer sets the joint. Use perforated pitch-fiber pipes for septic seepage lines.

Comparison of popular sewer pipes			
Pipe	Advantages and disadvantages	Relative cost (PVC/ABS/RS=1)	Sizes available
PVC/ABS/RS plastic	Lightweight, good flow, resists corrosion, easiest to install. Joints cannot be undone in case of error. Few tools needed. Needs rock-free sub-base and fill, plus uniform support for its entire length. Wide selection of fittings. Adapts to other pipes. Resists root infiltration.	1.0	3", 4", 6" in 10' lengths
Pitch-fiber	Not heavy, will not corrode. Joints can be taken apart, but require tooling to use cutoffs. Easy to install. Comes in long lengths.	1.8	3", 4", 6" in 5', 8', 10' lengths
Clay tile	Hard, rigid, durable. Difficult to work with, cut, and join, without special self-sealing joints. Prone to root infiltration. Joints permit offsetting in trench without fittings. Few fittings available. Few tools required. Heavy, breakable, short lengths. Builds into municipal-quality system.	2.2	4", 6" in 1', 2', 5' lengths
Cast iron No-Hub	Strongest, most indestructible. Resists corrosion and rock damage. Adapts to other pipes. Joints may be taken apart at will. Easy to add onto. Difficult to buy, except from a plumber. Heavy to work with. Easy to install. Flexibility of joints allows offsettting without fittings. No root infiltration.	5.1	3", 4" in 5', 10' lengths

Adapting pipes

What can be adapted

Water supply
Female threads to copper sweat
Male threads to copper sweat
Female threads to flared copper
Male threads to flared copper
Female threads to garden hose bibb
Male threads to garden hose bibb
Female threads to rubber hose
Female threads to CPVC plastic
Male threads to CPVC plastic
Female threads to compression
Male threads to compression
Female threads to polyethylene
Male threads to polyethylene
Brass threads to galvanized threads (dielectric)
Copper sweat to galvanized threads (dielectric)
Copper sweat to CPVC plastic
CPVC plastic to PB plastic
Copper to PB plastic
CPVC plastic to faucet supply
Chromed brass to faucet supply

Drain-waste-vent
ABS/PVC DWV to caulk cast iron
ABS/PVC DWV to clay tile sewer
ABS/PVC DWV to pitch-fiber DWV
ABS/PVC DWV to ABS/PVC/RS sewer pipe
ABS/PVC DWV to copper DWV
ABS/PVC DWV to threaded DWV
ABS/PVC DWV to trap
Downspout to PVC DWV
Hub cast iron to No-Hub cast iron
Copper DWV to caulked cast iron
Copper DWV to threaded DWV
Copper DWV to trap
Threaded steel DWV to trap
Threaded steel DWV to caulked cast iron
DWV of all types to cleanout

Special
Dishwasher to fixture trap
Sump pump to discharge pipe

What cannot be adapted
Water supply to DWV piping
Water supply to sewer piping
Water supply to lawn sprinkler piping (except through a vacuum breaker)
Water supply to hydronic heating piping
DWV pipe to buried "drain" pipe
Sewer pipe to buried "drain" pipe

1. When joining steel to copper in a hard water area, use a special transition fitting with rubber washer to prevent dielectric action. Solder the copper adapter to its pipe, and tighten the threaded portion to its pipe, before joining the union. 2. In soft water areas this copper sweat-to-threads adapter will work, but make the soldered connection first. 3. CPVC adapter screws directly into threaded fitting. 4. Plastic fitting for adapting flexible polyethylene tubing to steel has stainless steel clamp to hold tubing.

Pipe adapters are not only the most useful of fittings, but they are frequently essential, and they come in just about any form you could want. They allow one type of pipe to be connected to another. Many come in straight form, but others come in the form of elbows, tees, wing elbows, reducers, and valves. Some are intended to prevent dielectric action between pipes made of dissimilar metals. In hard water areas, dielectric action can cause electrolytic corrosion, as when steel and copper are combined in a water-supply system. The same goes for a DWV system.

Installing an adapter requires that you make up two different kinds of pipe fitting. One end of the adapter installs like fittings for one type of pipe; the other end installs like fittings for the other. Thus, to connect an adapter you may have to use a pipe wrench on one end and a soldering torch on

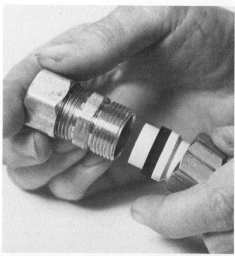

Adapters join different kinds of pipe, each pipe requiring its own joining method. Rubber washer on this transition adapter relieves stresses between the dissimilar materials—plastic and brass.

the other. Or, one end may be solvent-welded and one lead-and-oakum caulked.

All CPVC-to-copper-sweat transition fittings should be taken apart before you heat the solder side. Otherwise, the heat may destroy the plastic side and rubber mating gasket. For a male-threads transition, on which the collar cannot be removed for soldering, slide the collar 18 inches back along the copper tube. If this much tubing is not freely available, the collar and tube next to it should be wrapped in wet rags to protect them from heat.

A CPVC-to-male-threads transition fitting has insufficient space to tighten the male-threads portion into a threaded female fitting with a pipe wrench. A pair of channel-locking pliers—which have narrower jaws than a pipe wrench—will do the trick. Or, you can use a giant-sized Allen wrench on the broached inner hex nut.

Threads-to-sweat-copper adapters should be tightened at their threaded ends fully before soldering. Once soldered, they can no longer be threaded. Removal of them, if needed, requires heating the copper side to melt the solder and unthreading while still hot. Reinstallation will require heating and simultaneous threading. Apply more solder before leaving the joint.

Cleanout fittings, used in all drain-waste-vent systems, are wye adapters with a threaded cleanout plug. One type of PVC DWV piping uses an O-ring seal cleanout fitting hub. Every fitting thus becomes a cleanout adapter. The O-ring is first coated with petroleum jelly to ease installation.

A No-Hub joint becomes its own adapter in joining hubless cast-iron pipes to same-sized plastic, copper, clay and pitch-fiber pipes. Where sizes vary, a similar Calder coupling performs the same job.

Problems with low flow

Scale on the inside walls of galvanized steel pipes can sharply reduce water flow. Flow may be so feeble that it's hard to rinse dishes, and bathtubs take a long time to fill. Water running through the restricted pipe may make a loud running sound. The cure is new water-supply piping. Use flexible copper or polybutyline tubing which resists scale buildup and can be snaked around and through walls readily. If the low flow is caused by a combination of scaled-up pipes and low house water pressure (less than 40 psi), replacing the piping and installing a booster pump (see below) should cure the condition.

The only cure for scaled-up piping is to replace it. That's a big job, so tackle it in stages, perhaps doing the most affected runs first. Old pipes need not be removed first. Polybutylene pipes can be fished through holes, to save tearing out walls.

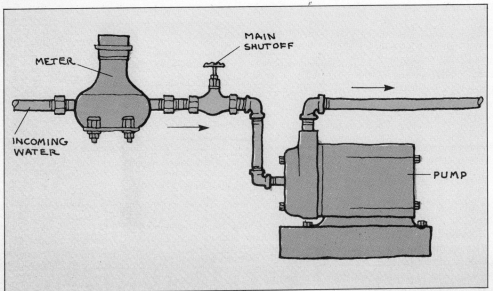
Pressure booster pump has an automatic switch that kicks on whenever a water tap is opened. It's expensive, but it will solve the problem of low flow from low house pressure if the water-supply pipes have unrestricted flow.

7. APPLIANCE REPAIR

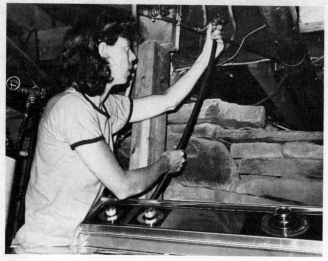
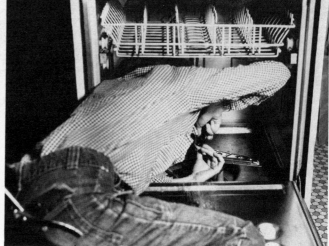

Clothes washers

What it takes

Approximate time: From a few seconds to several weekends. Set your own time limit. When you reach it, if you haven't corrected the trouble, call in a professional.

Tools and materials: Electrical test-light or volt-ohm-milliammeter (VOM), screwdriver, wrenches, pliers.

Planning hints: Repairs to automatic washers are often electrical or mechanical. If you have only a shaky knowledge of electricity, call a professional.

A modern clothes washer consists of electric motor, drive belt, transmission, water pump, tub, frame, and enclosure. It also has electrically operated hot/cold mixing valves. Controlling all of a washer's electrical functions—filling, shifting, agitating, pumping out, spinning—is a timer.

It helps to have a copy of the owner's sheet before you dig into a washer's innards. Areas you can probably tackle are water-inlet valve disassembly and cleaning, belt adjustments, pump cleaning, and checking for loose electrical connections. Do not try to tackle timer repairs or replacements, transmission or clutch work, or shifter repairs. As with all appliance repairs, pull the wall plug before touching any electrical parts. **Caution:** Electricity can be deadly. Never remove the access panel of a *plugged-in* washer. And don't ever poke anything, including your hands, into one that's connected. A professional, when making electrical tests, will plug the cord in, then, using one hand, cautiously touch the leads of a test light across portions of the washer's circuit. The other hand will be held behind to keep from creating a grounded path through the body. Electrical testing finished, the machine is unplugged again.

Troubleshooting an automatic clothes washer

Problem	Solution	What it takes
Washer won't run.	Make sure plug is in, fuse or circuit breaker is not out, both faucets are turned on, timer control knob is "on," load is balanced and washer door is closed. If it still won't run, call pro.	No tools required.
Will not fill.	Water valve off—turn on. Mixing valves not opening—check for loose lead, repair or replace valve. Pinched hose—unkink it. Sediment in hose inlet screen—clean. Bad timer—call pro.	Electrical test light.
Water won't shut off.	Valve stuck open from sediment or defect—clean or replace. Defective timer—call for service.	Test light.
Wash cycle not working.	Loose or broken drive belt—adjust tension or replace. Motor won't run—lighten load, reduce suds, check for loose leads, call for service. Thermal protection needs resetting—push red button in until it clicks. Defective timer—call for service. Also check for broken or loose wires.	Test light, wrenches.
Will not spin dry.	Uneven load—reposition. Broken or loose belt, defective timer, broken wire or loose connection—see same problems above. Broken transmission spring, brakes not releasing, defective bearings, motor not reversing if it should—call for service. Tub not drained—repeat drain cycle, unclog stopped drain, or check belt tension and pump (see below).	Test light, wrenches.
Will not pump out tub.	Broken or loose belt—tighten or replace. Clogged or locked up pump—clean or replace. Clogged or kinked drain lines—clean and straighten out. Timer or motor problems—call for service. Also, look for loose electrical leads or broken wires.	Test light, wrenches.

Dishwasher repairs

A dishwasher consists of tub, rack, motor with pump and impeller or spray arm, inlet valve, and Calrod heating element. A timer controls all electrical functions.

In the normal cycle, water enters tub, streams of water shoot against dishes, rinse water is pumped out and wash water enters. Sudsy water jets onto dishes and pumps out. This is followed by three rinses and the drying cycle.

Serious dishwasher repairs are best left to a professional. But you may try replacing a leaking drain hose or a worn-out door switch yourself.

What it takes

Approximate time: Three to four hours should be enough to troubleshoot most problems.

Tools and materials: Electrical test-light, pliers, screwdriver, Phillips screwdriver, wrenches.

Planning hints: While repairs often can be made with the dishwasher in place, you may want to work with the tub out on the floor.

Water flow during wash cycle

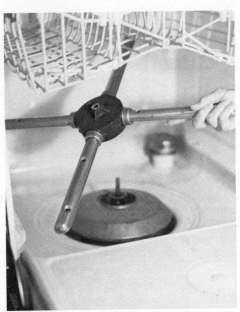

When dishes are not cleaned satisfactorily, the spray arm may be blocked. Make sure nothing is keeping it from rotating, and lift it out and clean out the holes in it. On some models, the nut holding it must be loosened before the arm will lift out.

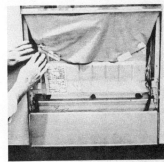

If condensation forms and runs down behind the door, install a vapor barrier. Using metallic tape, fasten it at the top. Tape the lower part to the door panel as shown.

Troubleshooting a dishwasher

Problem	Solution	What it takes
Washer won't run.	Make sure machine has power, door is latched, and cycle button is fully activated. If machine is between functions, wait for change. If utensil is blocking pump or impeller, remove. Thermal protection device may have tripped—push red button until it clicks. If none of the above work, call a professional.	Electrical test light.
Water will not enter.	Solenoid inlet valve defective—check for loose terminal or broken wire, remove and clean valve. Door interlock switch not working—adjust or replace switch. Overflow switch defective—test, replace if necessary. Filter screen completely clogged—remove and clean.	Test light, screwdriver, pliers.
Water won't shut off.	Sediment in solenoid inlet valve—take apart and clean. Replace needle, spring, and diaphragm if worn or damaged. Inlet solenoid problems—replace valve if defective. Faulty timer—call for service.	Test light, screwdriver, pliers.
Water stays in tub.	Motor won't reverse—call for service. Outlet strainer clogged—take apart and clean. Check discharge tube for blockage, kinking. Impeller problems—if jammed or sheared, repair or replace (may be a pro job). Timer or motor problems—call a pro.	Test light, screwdriver, wrenches.
Timer does not advance.	Call in a professional.	No tools required.
Incomplete drying.	Cool water—check water temperature. Should be 160 degrees Fahrenheit. Increase temperature setting, insulate piping to dishwasher. Calrod heating element defective—check at terminals (carefully) for current, replace if open-circuited. Call for repairs if timer is defective.	Test light, wrenches, thermometer.
Leaks from tub.	Door not closing on gasket—adjust, if possible. Hoses faulty—tighten clamps, unkink hose. At water inlet—tighten inlet fitting. Hole in tub—seal with silicone rubber or epoxy glue. Broken spray arm—replace. Gasket problems—install new gasket, or repair old one.	Screwdriver, pliers, glue, gasket.

APPLIANCE REPAIR 59

Sump pumps

What it takes

Approximate time: The time to troubleshoot a dead-in-the-water sump pump and get it pumping again can vary from half a minute to a good part of a day.

Tools and materials: A screwdriver, a pair of pipe wrenches, possibly a mechanic's wrench set for pump repair rather than replacement. A new float switch can be installed with a wrench and perhaps a screwdriver.

Planning hints: Don't tear into a working sump pump when rain is forecast. But if the pump does not work, you'd best get it going again before a storm comes. While a sump that is connected to footing drains will take lots of water before it overflows onto the basement floor, a good rain can provide that amount of water in minutes. Don't buy a new pump until you've determined that the old one cannot be rejuvenated with a switch repair or replacement.

Caution: A fatal shock can be suffered while working on a pump with an ungrounded internal short. Always unplug the pump (and the switch lead, if the unit has separate cords) before touching the pump or its metal piping. Also, make sure that the pump is grounded through a three-prong grounding plug used with a three-hole grounding receptacle.

A sump pump consists of a motor, a pump unit with impeller and volute-shaped housing, and a shaft connecting the two. In an upright sump pump, a long shaft is required to reach from the motor above the sump to the impeller at the bottom of the sump pit. A costlier, submersible pump has a very short shaft which is connected almost directly to the impeller.

Most sump-pump problems are electrical—chiefly corrosion of the switching mechanism. These can be corrected by repairing or replacing the switch. Other no-run problems come from mechanical switch failure, waterlogged floats, sticking actuation rods, and float supports hung up on internal pump or discharge piping parts. Another common sump-pump problem is plain wear-out of the pump impeller or, less often, a shot motor.

In some uses, sump pumps move significant amounts of sand and other gritty materials, causing rapid impeller wear. Furthermore, impeller bearings are water-lubricated, and when a pump runs dry and does not shut off, the bearing wears quickly.

Sump-pump maintenance consists of a regular cleaning of the sump pit—a not-so-pleasant job. With a tin can on a stick, scoop out debris that collects in the sump. Then spray the pit vigorously with a garden hose to stir up remaining debris while the pump runs to pump it out.

If the pump does not turn on when the water reaches its high level in the pit, or does not shut off before the level drops to the pump's intake screen, the problem is likely in the switch. Unplug the pump before investigating switch troubles. Pump-parts lists include new floats and new microswitch units. Installation of one of these may be the cure. But if the pump makes lots of noise when it runs, the bearings have probably worn out. While new bearings can be installed, replacement of the entire pump is usually more economical in the long run. The photo series on the next page illustrates this procedure. It is also possible to replace just the pump portion of most sump pumps. One type simply twists out, leaving the pump and motor separated.

Whether to replace with an upright or a submersible sump pump is a hard choice. The submersible costs more, but it will not be damaged by basement flooding and can safely pump it dry once power is restored.

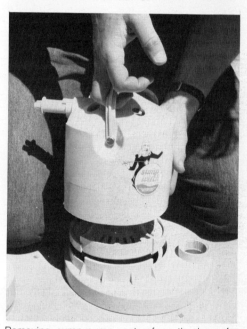
Removing sump-pump motor from the base for replacement.

New wrinkles

A new switch is available as original equipment or as a replacement for a worn-out sump-pump switch. The microswitch heart of the unit cannot become corroded because it is hermetically sealed inside a noncorroding vinyl float chamber. (Photos at right show float chamber cutaway to illustrate how it works.) The pump plugs into a piggy-back switch receptacle that itself is plugged into the wall outlet. In this way pump operation can be checked manually by plugging the motor lead directly into the wall outlet, bypassing the switch.

The new sump-pump starts running when its rolling steel ball drops into contact with the internal microswitch arm. Then, as the water level pumps down

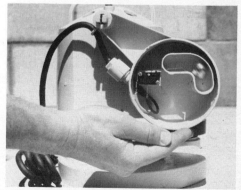
and the sealed floating switch chamber follows, the rolling ball moves out of contact with the microswitch arm, turning off the pump motor.

Replacing a sump pump

1. After removing the worn-out pump and any defective piping discharge, lower the new pump into its floor crock. It should rest level in the crock with the switch farthest from the sump wall. Properly positioned, the discharge pipe comes up in the center of the crock.

2. If you convert from a submersible to an upright model, you'll have to make a new crock cover. This one is ready-made of plastic with preformed cut-outs that need only to be sliced through.

How a sump pump works

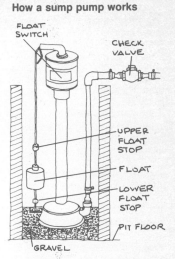

Water rising in pit pushes float up to the upper float stop which turns on the pump motor. As water is pumped out, float drops down to lower float stop which turns off the motor.

3. The sump pump's cover can be split and installed around the pump. If you use a plywood cover, make it of ¾-inch exterior-type plywood and cut in half to fit around the pump.

4. Every sump-pump installation should have a check-valve to prevent drain-back of discharge water with resultant short-cycling of the pump and waste of electricity. This check-valve is designed especially for sump-pump use and allows connection to vari-sized pipes.

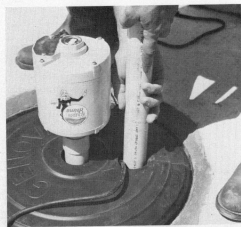

5. Run the discharge piping to a convenient location where sump water can be drained. Although harder to install, rigid pipe is preferred because it won't bend and interfere with the pump's float switch. Use 1¼-inch or 1½-inch pipe.

6. The last step is plugging the pump into a nearby electrical outlet. Be sure that it's a three-pronged grounding-type outlet. The pump features separate motor and switch leads. The motor plugs into a piggy-back receptacle on the switch lead.

APPLIANCE REPAIR 61

Dehumidifiers

Labels on diagram: CONDENSER, DEHUMIDIFIER COILS, HUMID AIR IN, AIR CIRCULATING FAN, DRY AIR OUT, SEALED REFRIGERATION UNIT, DRAIN BUCKET

What it takes

Approximate time: It takes only minutes to unpack and set up a dehumidifier for action. Maintenance is a pittance (see text).

Tools and materials: Unless your unit comes knocked-down, you won't need any tools to get it going. Mounting a humidistat on the wall will require a screwdriver.

Planning hints: A dehumidifier must be used in an enclosed space. So little as an open window nullifies its effectiveness, because vapor pressure from the more humid outdoor air will move humidity like steam into the drier room. Vapor can even move through walls. Walls, floors, and ceilings of rooms to be dehumidified should be protected with vapor barriers. Ordinary insulation batts and blankets contain vapor barriers. Paint or prefinished wall paneling and floor tile are also effective. Wallpaper, unpainted wallboard, plasterboard and the like are poor barriers, though they help. The surest vapor barrier is a layer of plastic sheeting placed over the studs before installing the wall material.

Rooms that are troubled with dampness and mildew make good prospects for dehumidification. A basement office or rec room, an upstairs den, a workshop where tools tend to rust because of high humidity during the off-season—all these can be dried out with a dehumidifier.

A dehumidifier works like a small air conditioner. However, while an air conditioner carries heat out of a room and deposits it outdoors, a dehumidifier puts the heat right back into room air. At the front of the machine is an evaporator, which is cold while the dehumidifier runs. In the center is a motor, fan, and refrigeration compressor unit. At the rear is a condenser. All three are plumbed together.

As air flows in through the finned coils of the cold evaporator, it cools rapidly, giving up much of its moisture. The effect is the same as when damp air meets a cold water pipe. Condensate drips off the evaporator into a collection container below. Cooled air then is blown out through the fins of the warm condenser to be recirculated. With its moisture thus wrung out, the air is more comfortable. An average dehumidifier can suck from the air up to 25 pints of water a day. However, air temperature, because of the unit's 500-watt-or-so power draw, may be several degrees higher than without the dehumidifier. But if the old maxim can be believed—that it's not the heat, it's the humidity—then the slightly higher mercury won't be noticed.

The cost of a unit varies according to how automatic it is. Top of the line is a unit that comes with a *humidistat*. This electric control has strands of horsehair that shrink and stretch along with changes in relative humidity. The action opens and closes a pair of electrical contacts. Wired through the contacts, the dehumidifier switches on or off. A reasonable setting for a humidistat, which is adjustable, is 50 percent relative humidity. If you set one too dry, the dehumidifier will run constantly trying to catch up.

Another method of humidity control is by means of a timer, but it has the drawback that as outdoor humidity changes, indoor relative humidity will change too, unless the timer setting is corrected.

One dehumidifier will handle one or two enclosed rooms. For a whole house you would need two or three large units. Units are rated by their water-removing capacity in pints per day at 80 degrees Fahrenheit and 60 percent relative humidity.

Dehumidification is needed in a non-air-conditioned house only during the heating off-season when indoor relative humidity climbs to about 60 percent. During the heating season you can pack the unit up and store it away.

Maintaining a dehumidifier

Costlier dehumidifiers turn themselves off when their drain containers fill. Otherwise collected water would begin overflowing onto the floor. Thus a dehumidifying unit that is not direct-draining needs daily emptying of its container. The iron-free water is excellent for steam irons and topping off the cells of auto batteries. The only other care a unit needs is once-a-month cleaning of the evaporator coils. To do it, the front panel of the unit is removed to expose the fins. A stiff shop brush or a vacuum-cleaner-brush attachment will remove built-up dust and dirt. After about a year of use, the unit's condenser coils may need cleaning, too. Any other problems—rare with a dehumidifier—require the service of a trained refrigeration technician.

Caring for your water heater

A booklet that comes with every water heater tells what steps are necessary to maintain it. Care is standard among the different makes of heaters.

Every modern water heater has what is called an energy-cutoff device. This is a nonadjustable temperature-limit switch that stops all energy flow to the heater should its temperature reach that limit. It prevents boiling of tank water. If the heater's temperature selector fails, the energy cutoff switch opens, stopping the heating process. Even though the cutoff closes again upon cooling, the pilot of a gas-fired burner goes out. It and the burner stay out. Continued pilot light outages preceded by higher-than-normal water temperatures indicate heater thermostat failure. A dealer correction is called for.

Some gas water heaters have air adjustments to obtain a blue flame. To adjust one, loosen the shutter locknut, raise the temperature setting to operate the burner, and rotate the air shutter to get a blue flame. Flame adjustment of oil-fired water heaters should be made by a service technician with combustion-checking equipment.

Every few months, drain a pail of water from your water heater. This gets rid of heat-insulating sediment at the bottom of the tank. Stop draining when the water runs clear.

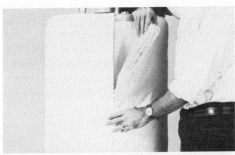

An additional layer of insulation installed around the outside of your water heater's storage tank will pay for itself rapidly at the present high cost of energy.

Stopgaps and copouts

You can save on your energy bill simply by setting back the heater's temperature selector one mark (20 degrees). If you overdo it, the last shower of the day may end up cold.

Check that relief valve!

For the safety of your household, the relief valve atop your water heater should be checked once a month for proper operation. If this valve gets stuck, its vital relief functions could fail. You'd be unprotected against explosion of a runaway water heater. Lift the pressure-test handle on the valve and run a few pints of water down the relief tube to be certain that the valve is working. It is a second line of defense against tank explosion.

Troubleshooting water heaters

Problem	Solution	What it takes
Insufficient hot water or water too cool.	Undersized heater—get a larger one. Too low a temperature setting—increase it. Defective thermostat—replace it. Bare piping—insulate it (see page 42). Heating element faulty—replace.*	Plumbing tools, pipe insulation, test light.
Water too hot.	Thermostat set too high—lower it. Thermostat defective—replace it.*	Test-light, screwdriver.
Boiling sound from tank.	Droplets of water beneath sediment turning to steam—flush out tank. Electric heater with scaled-up element—remove element and clean off scale or replace it.	Screwdriver, garden hose.
Rusty hot water.	Silt or mud in tank—turn off heater and water, turn on hot water and drain tank. Refill and flush. Turn everything on again.	Garden hose.
Popping relief valve.	Large water flow—check for too-high temperature setting or a failed relief valve.	No tools required.
High fuel consumption.	Scale in tank or around heating element. Improper air intake to oil or gas burner. Gas pilot too high—turn adjustment screw clockwise while watching flame. Uninsulated piping.	Screwdriver for pilot light.
Rapid tank corrosion.	Screw out the magnesium anode inside the heater tank every few years and inspect it. Install a new anode if it has been eaten away.	Screwdriver, wrenches.
Gas pilot snuffs out.	Install draft diverter if missing. Replace pilot thermocouple. Also, see text above.	Screwdriver.
Electric shocks.	Shorted element or thermostat—replace. Broken grounding wire—replace it. Defective insulation on lead wire—install new lead.	Test-light, screwdriver.

*Unless you're at home with electrical testing and wiring, troubleshooting of electric heater wiring problems should be left to competent service personnel. Do not take chances. Electricity can kill.

Water treatment

What it takes

Approximate time: Hooking up a water softener is a pleasant one-day project if everything goes well. Count as pluses: basement, garage, or utility-room water entrance. Count as potential problems: crawl space water entrance and attic water-supply mains. Copper and plastic water supply tubes make for the simplest softener installation; threaded pipes are harder.

Tools and materials: You'll need a full set of joining tools for your house piping. Or else plan on using short-cut adapter fittings that are available for galvanized steel piping and copper tubing. Materials include the necessary teeing-off fittings, piping to and from the softener, and connections to the treatment unit. One more thing you'll need is a drain hose leading to a place where you can dump regeneration water. In most cases this goes to a nearby floor drain or sink trap with a side connection for the hose. The hose may be rubber garden hose or polyethylene pipe with the necessary attachments.

Planning hints: A valid water test should precede your selection of a water treatment unit. If you use municipal or utility water, the water company will furnish an analysis of it. Beyond that, local treatment unit dealers probably know what's required in your area. If you draw water from a private water source, you'll need to make a water test to tell exactly what kind of treatment your water needs and how much treatment capacity is required. Capacity also depends on how much water you use. This can be estimated from the size of your family, the number of animals you have, and what other uses of treated water you will make. Plan on doing the installation at a time when the whole house water supply can be turned off for a few hours.

When the subject of water treatment comes up, most people think of a water softener. Yet softeners are only one form of water treatment. You can get a treatment unit to cure almost any water defect. The problem is that water on or under the ground contains many of the impurities of the ground: minerals, acids, and bases. These create hardness, stain fixtures, and smell and look badly. Treatment removes these unwanted products, making water more usable.

The primary problem is hardness, in which ground-rock transfers calcium and magnesium ions (electrically charged atoms) to the water. These ions react poorly with soaps and detergents, and so the water is considered hard. The modern water softener contains an ion-exchange resin bed that trades its good-acting sodium ions for bad-acting calcium and magnesium ions in the water. So, dissolved minerals are not actually removed, they are merely exchanged for more desirable ones.

When the resin bed becomes loaded with bad-acting ions and gives up most of its sodium ions, the process must be reversed, or else softening would come to a halt. Reverse processing is called regeneration. During regeneration, the resin is first given a rapid backflow—backwashed—to lift out collected solids. Then, switched to forward flow again, the unit resin bed is inundated with very salty water created by dissolving softener-salt in water and flowing it slowly past the bed. Sodium ions from the salt drop off into the bed and take the place of the more loosely attached magnesium and calcium ions. These go back into the water and are carried away to drain. Regeneration is, unfortunately, somewhat wasteful of water. This process—a required one for water softening—is one of the main drawbacks of home water softeners.

The simplest water-softener installation is to tap into the cold water main and connect it as shown at right. A good location is along the cold-water line between the last tap-off to an outdoor hose outlet and the first to a fixture other than the toilet. If just hot water is to be softened, connect only into the cold-water line going into the heater (see opposite).

Turn off the main shutoff valve and locate the mineral tank under the pipe it will connect to. Measure perpendicularly from the tank inlet and outlet openings up to the pipe and mark the pipe. Then, allowing for the makeup space required for the two adapters and elbows shown, cut out a section of the pipe and attach the adapters

How a water softener works

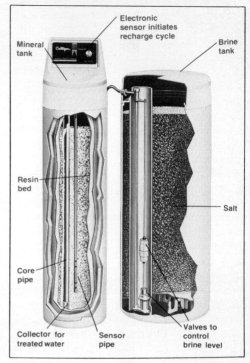

Running water through the resin bed converts calcium and magnesium ions—which react badly with detergents—into good-acting sodium ions. Resin is periodically backwashed with brine to regenerate its ion-exchange ability.

Water-softener connections

64 APPLIANCE REPAIR

and elbows to the pipe ends. Then connect flexible riser pipes from the tank's inlet and outlet connections to the elbows.

For an installation that lets you bypass or remove the softener when you want to, install the two shutoff valves and the bypass

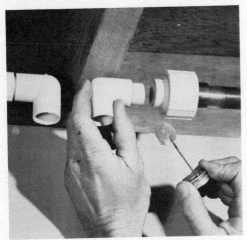

valve shown below. Softener installation kits come with many of the pipes and fittings needed. Fill with softener salt and you're in business.

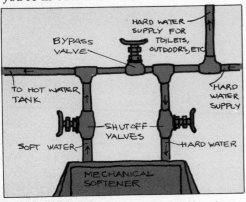

To soften water, both tee-off valves should be fully open and the main valve between them should be closed. This makes water pass from the main through the softener and back to the main again. Modern softeners also contain built-in bypasses that conduct hard water around the softener during regeneration. If, for any reason, the softener must be taken out, the manual bypass you have created with your three valves will provide hard water to the whole house.

If other types of treatment units are required, all should be located ahead of the water softener. All, that is, except a small activated-carbon filter placed in the cold water line and connected to a fixture where your drinking water is drawn. These filters remove scum and suspended solids and

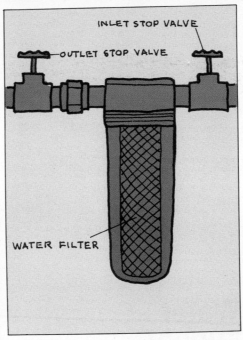

counteract bad taste in the water you drink. They're good at removing chlorine, too.

Many water treatment units are said to "purify" water. Do not make the mistake of thinking that they remove harmful bacteria. Only a chlorinator can do this by killing it with small amounts of chlorine. If your drinking water won't pass a health department purity test, then it needs germ purification before it can be safely drunk.

Benefits of a softener
Savings on soap and detergents.
Cleaner clothes.
Longer-lasting clothes.
Cleaner skin.
More lustrous hair.
Tastier foods.
Sparkling dishes.
Prevention of scale in water pipes.
Longer life for appliances.
Cleaner-looking fixtures.

Arrangement of water-treatment units

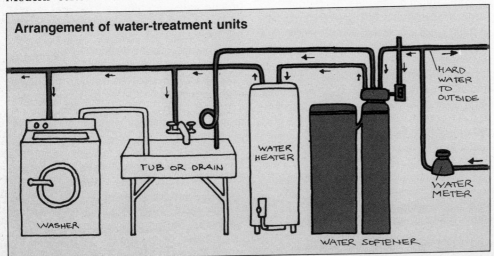

8. PRIVATE SYSTEMS

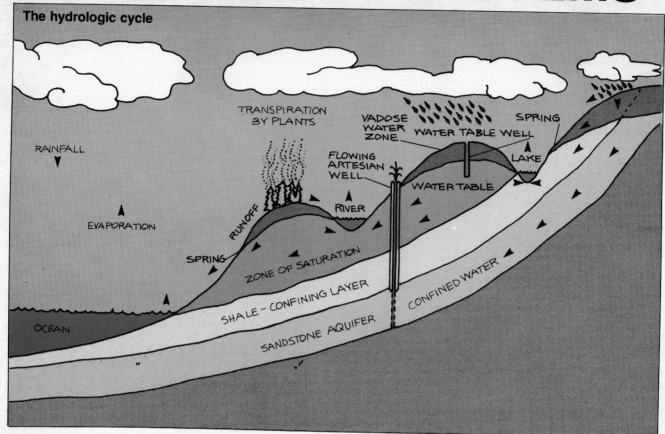

The hydrologic cycle

Water supply

What it takes

Approximate time: It takes a professional well-driller about one week to drill an average water well. Hand-digging a shallow well could take months, but developing a nearby spring to produce household water could be done in a long weekend.

Tools and materials: Drilling your own deep water well is not recommended. To install a pump in a well, you'll need steel or polyethylene pipe and fittings, wrenches, screwdriver, pliers, the pump, electrical controls, well seal or pitless adapter, an electrical switch box and cable, and a pipe dog to keep the pipes from slipping into the well while you make connections.

Planning hints: If you don't fully understand the electrical side of pump hookup, hire an electrician to handle it, and if your well is much more than 150 feet deep, you may prefer to have the well-drilling crew handle the lowering of piping into it.

The water can come from several sources: rain or melting snow; a surface supply, such as a lake or river; and in-the-ground water. This last source—ground water—is the typical water source for a private supply system. Water-table ground water closest to the surface results from water saturating the ground. A water-table source is likely to be contaminated, and its depth tends to vary seasonally, depending on past rainfall. In most areas, water-table wells are second rate, if not altogether unusable, for drinking water.

Deep water trapped in what's called an *aquifer* beneath layers of rock is usually pure, though sometimes heavily mineralized. To get at it, a deep well must be drilled through the rock layer into porous strata beneath. (Be sure you own the water rights.)

A third kind of ground water is the spring or seep. It brings ground water from a higher elevation, sometimes moving it many miles, then flows it out onto the ground.

Vertical well systems. These consist of motor, pump, pressure tank, and the piping between them. An automatic pressure switch turns the pump on when pressure falls below a preset level. It turns the pump off when it has pumped back to the desired maximum pressure.

In some ways, a private water system is better than a public one. Water from a private source is chlorine-free. And there's no water bill to pay. But the higher electric bill for pumping may make up for the free water supply, and you must maintain the pumping equipment yourself.

Before you drink well water, it should be tested either by public health authorities or a private testing laboratory. This is required in many areas, and in some areas, a permit must be obtained to drill a private well.

Pumps for residential wells. There are four types: centrifugal, piston, jet, and submersible (see page 68). They may be shallow-well—where water level is within 25 feet of the pump—or deep-well. Centrifugal pumps are shallow-well only. The others may be used in shallow or deep wells.

Drilling a well

The most practical way to drill for water is to hire a professional. With modern equipment it takes only a few days to bore down under hundreds of feet through solid rock. Lining the hole with well-casing pipe prevents it from collapsing. Wells are usually cased down to solid rock, sometimes all the way from top to bottom.

Installing casing and piping

Lowering 12-gauge culvert pipe that will form the casing for a well 30-inches wide that was drilled with a 36-inch auger.

Grout cement is worked around the culvert pipe casing along its entire length to strengthen the well installation.

New wrinkles

For drilling wells in soft rock such as limestone, you can buy a do-it-yourself well-drilling outfit at a price less than you might have to pay for a single professionally-drilled well. The driller comes with power unit, drill pipe, and hardened drill bit. You supply water from a garden hose to flush out drillings. About 100 feet is a comfortable do-it-yourself drilling depth. Where the unit really pays off is in drilling multiple wells for yourself and your neighbors.

Submersible pump is attached to a length of discharge pipe and then lowered into the well casing. If the well is not too deep, the pipe may be polyethylene plastic. Installations over 220-feet deep should be made with galvanized-steel pipe.

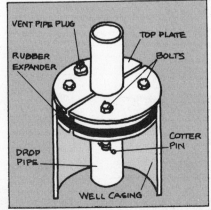

The well seal at the top of the casing prevents infiltration of surface water and other contaminants. Two pipes going into a well indicate a deep-well jet. A single pipe indicates that either a shallow-well jet or a submersible pump is in use.

Well-pump pros and cons

Jet pump: High volume, low energy consumption, few moving parts, simple to repair. Sediment or scale in water may lodge in jet and plug it. Can be tough to cure if there is much pipe to pull out of well. Flow rate drops drastically at greater depths.

Submersible well pump: Requires larger well. Deep-pumping models available with low pumping rates. Trouble-free pumping, except in sandy wells where they're not suited. Vulnerable to lightning strikes and to pumping well dry. The best, but the costliest residential-type well pump.

Shallow-well centrifugal pump: Simplest pump for low-pressure, low-lift applications. Low cost and practically trouble-free.

Piston pump: Can develop extremely high pressures. Most practical as a low-volume, shallow-well pump. Has more wearing parts than other pumps.

How a pressure tank works

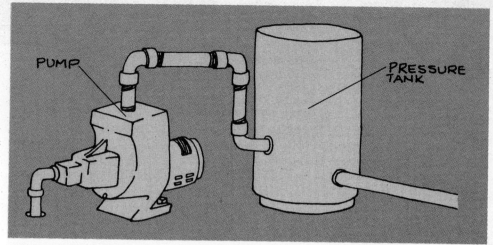

Well pipes lead to the pressure tank, which keeps water stored under pressure for use whenever a tap is opened. The pump motor will cycle on and off automatically at preset pressures.

Pumping depth vs. gallons per minute

Jet pump Pump horsepower	Pumping depth in feet, to yield given gpm at 40 psi												
	40	60	70	80	100	120	140	160	180	200	210	240	260	280
¾	8.4	5.8	4.7	4.3	3.7	2.7								
1	11.7	11.1	10.2	9.2	6.4	5.1	4.5	4.0	3.2	2.9	2.1			
1½	12.2	12.2	11.7	11.3	9.9	8.3	5.0	4.9	4.9	4.7	4.3	3.7	3.0	2.3

Submersible pump Pump horsepower	Pumping depth in feet, to yield given gpm at 40 psi									
	20	60	100	140	180	220	260	300	340	380	450
½	12.3	10.8	9.1								
¾		12.8	11.4	10.4	8.8						
1				12.1	11.1	10.2	9.4	8.0			
1½						12.1	11.6	11.0	10.3	9.3	7.4

Troubleshooting well pumps

Problem	Solution	What it takes
Pump will not run.	Check power, fuse, circuit breaker, switch, pressure switch, wires leading to motor, flow-cutoff switch (if used). If current reaches pump but unit doesn't run, call for service.	Test-light, VOM.
Pump draws power but doesn't pump.	Jet pump: debris in jet—remove and clean. Submersible pump: locked rotor—check amperage draw and compare with locked-rotor specifications in manual. Also, check water level below intake pipe, lost prime, or stuck check-valve.	Ammeter, pipe tools, pipe dog.
Pump runs briefly, then stops.	Pressure tank is waterlogged—recharge tank. Pressure switch differential set too close—adjust switch to increase differential to about 20 psi. Pressure switch installed too far from pressure tank—locate switch next to tank.	No tools required.
Pump delivers water too slowly.	Well-water level has receded—lower pipes deeper into well. Pump too small—check for valve sticking. Inadequate voltage to pump—check voltage with pump running and compare with manufacturer's minimum requirements; use heavier wires to increase voltage. Jet pump: partially plugged jet—remove and clean.	Pipe tools, well tools, new wires, VOM.
Pump starts when no water is being used.	Leaking check-valve—remove and replace. Other leaks: listen at well casing for a spraying sound—pull out pipes and repair leak.	Pipe tools, well tools, check-valve.
Pressure low or variable.	Adjust setting of pressure switch to correct. See the pump manufacturer's manual for recommended pressure settings. Switch has two settings, one for level of pressure, another for differential between lowest and highest pressure.	Screwdriver, pliers.
Noisy pump.	Motor bearing worn out—replace bearing.	Shop tools.
Leaks around pump shaft.	Seal worn out—replace.	Pipe tools.

Jet pumps are the most common for both shallow and deep wells. The pump's impeller spins water from center to outside. Discharge water flows into a pressure pipe leading to a 180-degree turn, through an orifice called a *jet*. The jet leads up to a larger suction pipe, and is open to well water at the bottom. The powerful jet of water streaming up the suction pipe carries some well water along with it. At the pump, water is forced back down through the jet. The excess flows off to the pressure tank for home use.

On a shallow-well jet pump, the jet is mounted on the pump and only a single suction pipe from the backside of the jet reaches into the well. Motor, pressure tank, and controls mount with the pump. For deep-well use, the jet is placed down in the well within 25 feet of the lowest water level.

Piston pumps use a piston in a cylinder, working like a syringe, to suck water up out of the well. For deep-well use, the piston is placed down in the well and worked by a motor-driven rod. The mass of moving parts makes it less desirable than a deep-well jet pump, unless wind-driven.

Centrifugal, jet, and piston pumps must be primed before they will pump. This is done by pouring water into a priming opening, then closing it and operating the pump. Normally, no further priming is necessary.

Submersible pumps have a long, narrow motor directly driving a long turbine pump with multiple impellers. Supported below the lowest well-water level by the discharge pipe, it may also have a safety cord to prevent pump loss down the well if the discharge pipe becomes detached. A waterproof cable goes down the well with the pump to bring it electricity. Controls and pressure tank are above ground.

In any well, it is important that the intake opening be low enough to always be under water. Air entry makes a pump lose its prime. In a submersible type, lack of water can even destroy the pump. An intake should not be at the very bottom of the well, however, where it could draw in debris. An experienced well-driller can usually tell you what is the best depth to place your pump parts and can accurately estimate what the pumping capacity will be.

The gallons per minute rate of a pump depend on the type of pump, pumping depth, friction pipe losses, and the pressure required at the farthest faucet. Well pumps come in a number of models (see table on opposite page) and well and pipe size should be integrated with the pump used.

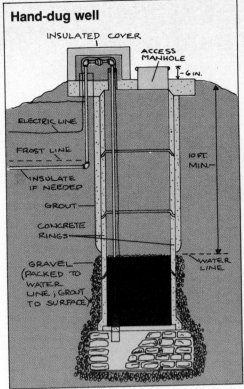

Hand-dug well

Wire sizes for pumps

To be certain submersible-pump motors have adequate voltage, the size (gauge) of cable serving them must be right. Figure the distance from switchbox to well, plus that from wellhead to submersible pump. Then consult the table below. Wire sizes are AWG copper.

Cable length	Cable gauge
½–hp motor (115-volt)	
To 35'	No. 14
55'	No. 12
85'	No. 10
140'	No. 8
220'	No. 6
½–hp motor (230-volt)	
To 140'	No. 14
220'	No. 12
340'	No. 10
560'	No. 8
875'	No. 6
1–hp motor (230-volt)	
To 85'	No. 14
140'	No. 12
200'	No. 10
340'	No. 8
525'	No. 6

Lightning bug

I didn't believe it when they advised me to install a lightning arrester on my submersible-pump control box. One night, lightning struck the wiring and grounded itself through my submersible pump. It was burned out, and I was burned up. It cost me a lot to replace the pump. I installed a low-cost lightning arrester which fits on the pump's control box. Now my pump is protected.

Practical Pete

Comparison of well types		
Type of well	Advantages/disadvantages	Cost ranking
Professionally drilled (to 1000' deep, 4"-24" diameter).	Most trouble-free. Can be any size, any depth, depending on ground factors. Best chance of getting bacteria-free water, but water may have high mineral content. Least dependent on ground water table. Can choose solid or perforated casing, as strata permit. May not need to be cased beyond first rock strata.	Highest.
Hand-dug (to 50' deep, 3"-20" diameter).	Hardest to make. Not workable where water table is far down. Best in porous soils away from sources of pollution and near a lake or stream for recharge. Greatly affected by seasonal water table. Flow may be low. Risk of collapse.	Low.
Hand-driven (to 50' deep, 1¼"-2" diameter).	Excellent where potable water exists above rock strata and ground is not rocky. Numerous wells can be driven vertically or horizontally to tap more water than one well can produce. May be hard work. Flow not as great as in a drilled well.	Lowest.

Sewage treatment

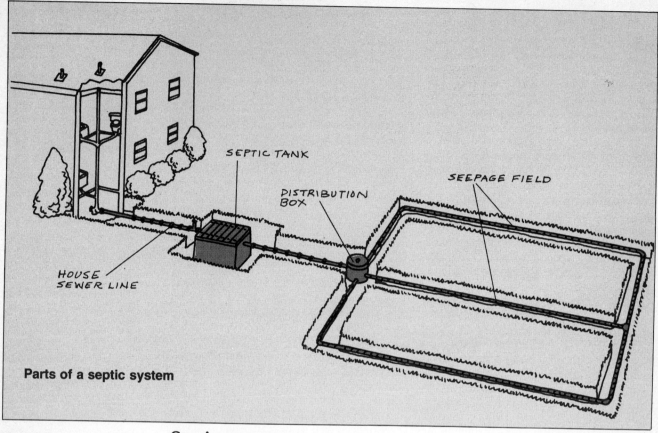

Parts of a septic system

Septic systems

Living in an area not served by a municipal sewage disposal plant requires that an alternate method be found for getting rid of sewage. The most commonly used is a private sewage-disposal facility called a septic system.

Once household wastes travel from the house sewer, they enter a large tank where the septic system takes over. In this tank, decomposition takes place by bacterial action. Solid wastes settle to the bottom and are pumped out every two or three years. The remaining liquid, called *effluent*, flows out of the tank through another sewer pipe into a distribution box which distributes effluent among seepage lines, where it seeps out through perforated pipes or loose-jointed tiles into the surrounding soil.

Septic tank sizes vary from about 500 gallons capacity to 1250 gallons or more. The larger the tank, the less frequently it needs to be maintained. Use the table on the next page to help estimate how large a septic tank you'll need.

To determine the rate at which your soil will absorb effluent, run a percolation test: Dig half a dozen or so holes, spaced evenly throughout the area where you plan to install the lines. Go about 36 inches deep; then put 2 inches of a porous material, such as gravel or coarse sand, into the bottom of each hole. For a minimum 4-hour period—better yet, 12 hours—saturate each hole by keeping it full of water. Now you're ready to make the test.

Refill each hole to a depth of 6 inches above the porous fill. After 30 minutes, measure the distance that the water level has dropped. Then refill the holes to 6 inches above the fill. Do this for eight 30-minute periods—a total of four hours. During the last period, measure the time it takes for the water level to drop 1 inch. Using this time as your key, refer to the Percolation Table, opposite, for the minimum amount of seepage area you should have for each bedroom in the house. Percolation rates shouldn't vary much from one test hole to another, but if they do, average the results.

For extremely porous soil that soaks up water in less than 30 minutes, use a somewhat different testing procedure: Allow only 10 minutes between refillings, and run your test for just one hour rather than four. The percolation rate then would be the time the soil needs, in the last 10-minute period, to pull down the water level 1 inch.

What it takes

Approximate time: How fast are you with a shovel? Soft, sandy soils may dig quickly; clay-type soils not so easily. Installing a septic system is just plain, hard labor. You may prefer to hire a machine and operator to do the actual digging.

Tools and materials: Saw, level, pipe-joining tools, sledge hammer, long straightedge. Trowel and mortar for some septic tanks. And don't forget the shovel and wheelbarrow.

Planning hints: Most pleasantly done during cool spring and fall weather. Digging frozen winter soil is impossible; sultry summer weather is unbearable. Ask your tank supplier for the necessary hole dimensions to accommodate your tank: go that big and no larger. Inspections may be required, so contact your local health department.

Building a sewer

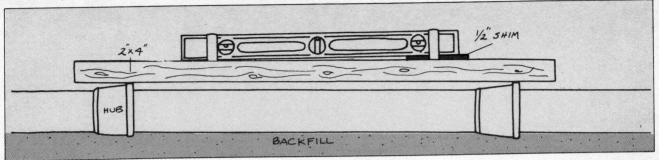

1. A sewer line should slope evenly away from a house at ¼ inch per foot. A 2-foot carpenter's level with a ½-inch plywood scrap under one end and taped to a long, straight 2x4 will help achieve slope. Sewer may slope steeper, but the last 10 feet should conform to the mentioned ratio.

2. Sewer pipes, except clay tile and cast iron pipe, need uniform support in the trench bottom. Dig out for pipe hubs so that pipes are supported along their full length. Trench bottom should be unexcavated earth, sand, or pea-gravel fill.

3. All rocks in contact with the pipes on trench bottom and in backfill should be removed. Earth backfill is okay, provided that it is fine and rock-free. Final fill above the pipes may be rocky.

Installing a tank

The easiest septic tank to buy is a precast concrete one. The dealer will tell you what size hole to dig. Be sure it's large enough, and that the bottom is perfectly level. Tank suppliers are equipped to deliver tanks in place. An A-frame on the rear of the truck lowers a tank into the hole facing in the direction you want it. When tank is in place, racks are removed.

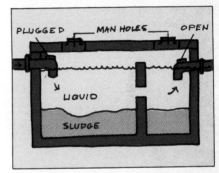

The typical two-compartment septic tank lets solids settle out at the bottom while sewage decomposes by bacterial action. Sewage enters through a plugged tee, leaves tank through a lower, open-topped tee.

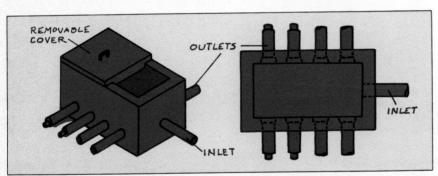

A distribution box accepts effluent from the septic tank and divides it equally among two or more seepage lines. Its outlets, which are all at the same elevation, sit slightly lower than the inlet. The distribution box itself must be level.

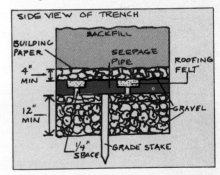

A typical seepage trench—12 to 36 inches wide, 3 feet deep—either slopes slightly away from tank or is level. The bottom 12 inches of trench are filled with coarse stones. Perforated pipes are laid and topped with more stones.

Percolation table									
Percolation rate: minutes required for water level to drop 1 inch after initial saturation.	2 or less	3	4	5	10	15	30	45	60	Over 60
Minimum seepage area: in square feet of trench bottom per bedroom.	85	100	115	125	165	190	250	300	350	Soil not suitable

Tank-size table				
Number of bedrooms	1–2	3	4	5
Minimum tank capacity	750 gal.	900 gal.	1000 gal.	1250 gal.

Building a seepage field

1. Position grade stakes in seepage trenches to get the proper pipe slope. For one of 2 to 4 inches per 100 feet, tape a 1/16-inch shim under one end of a 2-foot level. Position the shim at the lower stake, then tap down until bubble is centered.

2. Dump the initial stone fill into the trench bottom and level it off even with the tops of the grade stakes. Both the trench bottom and the fill should have the same slope.

3. Lay the seepage pipes on the gravel. Perforated pipes should be placed perforations down. Instead of solvent-welding the joints, use loose couplings. Cover joints of open-jointed clay tiles with strips of asphalt felt or shingles.

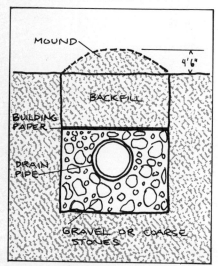

4. After covering the seepage lines with 4 inches of coarse stones, roll untreated building paper out over the stones. A layer of straw will do, also. Then back-fill with loose earth, mounded up above grade.

5. Ends of seepage runs need to be blocked off so effluent doesn't run out and erode the trench. Do it by placing a large rock against the pipe end. Or use a cap.

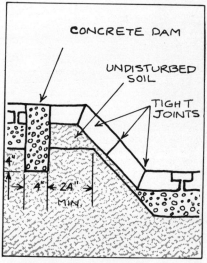

6. Changes in trench elevations call for cast concrete dams. Properly constructed, effluent in the higher trench cannot overflow into the lower trench without going through the pipe. Pipes between trenches are tight, not perforated.

Layouts for seepage fields

Allowable sludge accumulation
(bottom of outlet to top of sludge)

Tank capacity	Liquid depth			
	2½'	3'	4'	5'
750 gal.	5"	5"	10"	13"
900 gal.	4"	4"	7"	10"
1000 gal.	4"	4"	6"	8"

Smaller tanks require more frequent cleaning.

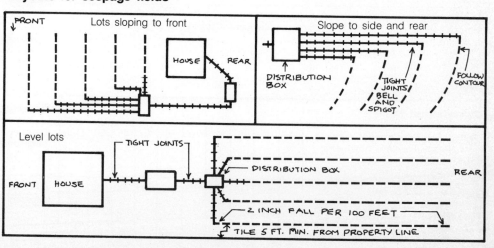

72 PRIVATE SYSTEMS

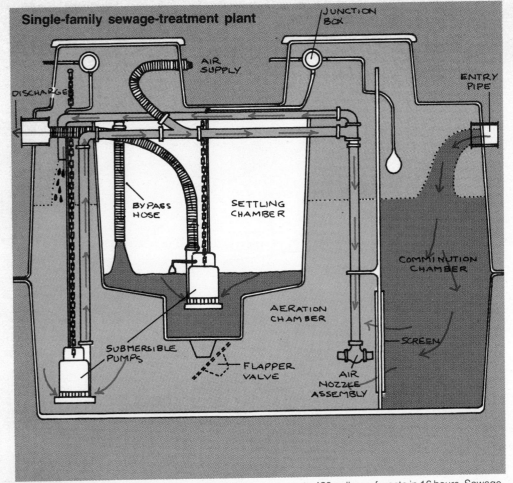

This aerobic sewage treatment unit for homes can process up to 480 gallons of waste in 16 hours. Sewage flows through entry pipe into communion chamber. As it flows from there through screen into aeration chamber, air from nozzle, drawn in through air-supply hose, breaks up and mixes the waste materials. Floating materials flow into the settling chamber and are aerated by air from the bypass hose. Treated effluent is then pumped out through the discharge pipe. Any settled solids are voided back into the aeration chamber through the flapper valve. Process is noiseless and odorless.

What it takes

Approximate time: A little longer than a septic system, so you may prefer to hire a contractor. Additional time is needed for the electrical wiring that serves the pumps. Digging and seepage field work (unless discharge is into a stream) will take about the same time as a septic system.

Tools and materials: Calls for the same tools as required for a septic system. If you do the wiring part of the job yourself, you'll need wire-working tools, too.

Planning hints: Work closely with both the dealer who sells the unit and local health officials. If you can, get the two parties together to advise you.

Stopgaps and copouts

If your ground is unfit for a conventional seepage field, a dug-out seepage pit with walls of loose-jointed concrete blocks can be put in. No bottom is installed other than a foot-thick layer of gravel to allow effluent to percolate into the soil.

A seepage pit is usually called for when an impermeable surface layer of soil sits atop a permeable layer, or when the available ground is too hilly to accommodate seepage lines.

In some areas, a precast concrete seepage pit, substituting for the home-built block one, can be purchased. If a single pit won't handle the load, a series of them, served by a distribution box, may be used. Not all codes okay seepage pits, so be sure to check before you start digging.

A pit should be located at least 100 feet from any water supply source, 10 feet from property lines, and 20 feet from buildings. In multiple-pit installations, the distance between them should be at least three times the diameter of the largest one. Pit diameter should be 4 feet or more and dug into at least 4 feet of porous soil.

If you build the pit with special loose-laid seepage-pit blocks, lay them with their small openings facing into the pit and their large openings facing out. Back the wall materials with at least a 3-inch thickness of coarse, crushed stones up to the level of the inlet. Above the inlet, lay the masonry with solid mortar joints.

Sewage-treatment plants

A step above the septic system for handling household sewage is the private sewage-treatment plant, a number of types of which are available. Select a system that is locally approved or one for which you can secure an experimental installation permit.

The sewage-treatment plant eliminates some of the drawbacks of the anaerobic (without air) septic system, chiefly the quality of effluent. Liquid left over after a septic system's 70-day bacterial-action process is finished is unhealthy and smells bad. Septic effluent cannot be discharged onto the ground, or into a stream or lake. It is so high in biological-oxygen demand that it would kill plant and animal life.

Fully treated 4-hour effluent from a private aerobic (with air) treatment plant is the next thing to being pure enough to drink. After chlorination, some health departments will permit its discharge into a flowing stream. Thus, if you're saddled with soil that is poorly suited for a septic-tank seepage field, the private treatment plant could be your answer. Even when used with a seepage field for disposing of effluent, a sewage-treatment system beats a septic tank. Septic effluent may, in time, fill soil pores with suspended solids—especially if the tank is not regularly maintained. This clogs up the field, making it necessary to extend it into a new seepage area.

Treatment-plant effluent is free of suspended solids. It will not harm a seepage field any more than plain water would. Also, if a seepage field should become waterlogged, treatment-plant effluent is not as objectionable to neighbors or passersby.

The cost of a private sewage-treatment plant is about twice that of a simpler septic system. Some units come ready to bury or to install aboveground in a shed; others need cast-in-place concrete tanks.

PRIVATE SYSTEMS 73

9. HEATING

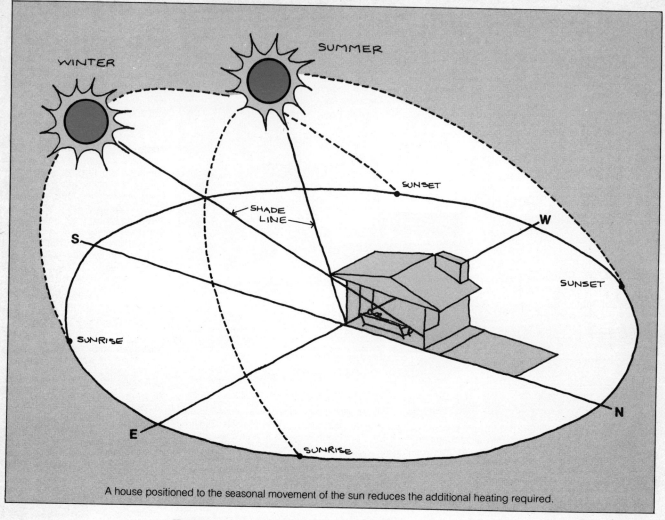

A house positioned to the seasonal movement of the sun reduces the additional heating required.

Energy shortages and increasingly higher fuel costs have prompted homeowners to take a closer look at home-heating systems.

This chapter will help you understand how various heating systems work and how to get the most heat with the least fuel consumption. Each section includes tips on how to maintain and service your heating system.

Hot and cold, warm and chilly, are relative terms. An electric heater may be hotter than a radiator, but colder than a furnace. In most parts of the country, an outside temperature of 50 degrees is warm for January, but cool for July. Differences in temperature cause heating systems to work. Heat always moves toward a cooler area. This movement of heat makes it possible to generate heat in one part of the house and then move it to cooler areas.

Heat moves in three ways

All forms of heat movement can, and usually do, occur at the same time. The various types of heating systems are designed to cause one kind of movement to predominate.

The purpose of a home-heating system is, of course, to maintain a comfortable temperature in the house. But comfort is dependent on humidity as well as temperature. Humidity is moisture content in air.

In warm weather, when the humidity is high, the air is holding as much, or almost as much, moisture as it can. Our body's cooling system depends on water evaporation (perspiration) to lower body temperature. The body's cooling system doesn't work well in a humid area. The moisture stays with us and we feel "sticky." In cool weather, the reverse occurs. The chilliness we feel when humidity is low (air is dry) is caused by rapid evaporation. So, both humidity control and temperature control are important in maintaining a comfortable home temperature (see opposite page).

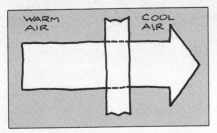

Conduction: In winter, heat is conducted through the walls toward cooler outside air, and the house loses heat. In summer, the reverse happens. Storm windows, insulation, reflective surfaces all slow down the heat conduction loss.

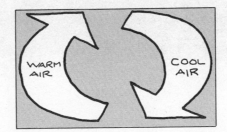

Convection: Warm air is lighter and less dense than cool air. This makes warm air rise and cool air descend, a movement called convection. It can cause annoying drafts, or help to keep temperatures even, if controlled.

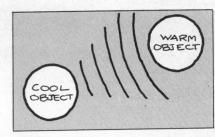

Radiation: Warm objects radiate heat toward cooler objects, as the sun warms the earth or as a radiator warms the people in a room.

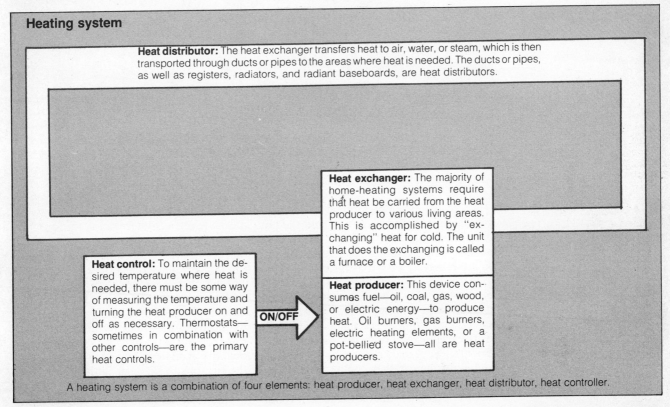

Heating system

Heat distributor: The heat exchanger transfers heat to air, water, or steam, which is then transported through ducts or pipes to the areas where heat is needed. The ducts or pipes, as well as registers, radiators, and radiant baseboards, are heat distributors.

Heat exchanger: The majority of home-heating systems require that heat be carried from the heat producer to various living areas. This is accomplished by "exchanging" heat for cold. The unit that does the exchanging is called a furnace or a boiler.

Heat control: To maintain the desired temperature where heat is needed, there must be some way of measuring the temperature and turning the heat producer on and off as necessary. Thermostats—sometimes in combination with other controls—are the primary heat controls.

Heat producer: This device consumes fuel—oil, coal, gas, wood, or electric energy—to produce heat. Oil burners, gas burners, electric heating elements, or a pot-bellied stove—all are heat producers.

A heating system is a combination of four elements: heat producer, heat exchanger, heat distributor, heat controller.

Humidity

Correct level: What is the correct level of humidity? The most practical indoor humidity level is between 30 and 35%. If the humidity is too high (40% or more), a significant amount of condensation can form, possibly even resulting in serious damage.

Control: Normal household activities such as washing, showering, and cooking add humidity to the air. Normally this provides 10 to 15% relative humidity. To reach and maintain the desired level, a humidifier must be installed in the house.

Humidifiers can be readily added to hot-air heating systems. The humidifier can be installed in the air ducts, close to the place where heated air leaves the furnace. A humidistat, built into the unit, controls the humidity level. A plumbing connection is required to supply water to the humidifier.

With any other type of heating system, a separate humidifier is required. These units require both electrical and plumbing connections. For best results, humidifiers should be centrally located in your home. For proper operation, doors should be left open as much as possible throughout the house.

Before purchasing a humidifier, you need to know the volume of living space and lowest expected outdoor temperature. To calculate the volume (in cubic feet), multiply the floor area of each room by the height of the ceiling. Add up the volume of all the rooms, hallways, and entrance ways to get the total volume. Include the basement and attic unless they are separated from the living area by insulation with a vapor barrier.

Calculated trouble
When I bought a new humidifier, I did not calculate *both* the volume of living space *and* the lowest expected outdoor temperature. Now that my new wallpaper has started to peel off, I know what I did wrong. Next time, I'll be sure to double-check all my calculations with the data supplied by the manufacturer.
Practical Pete

Heat producers

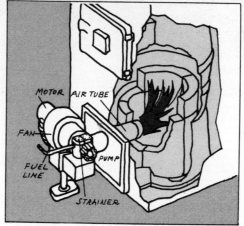
Pressure-type oil burner

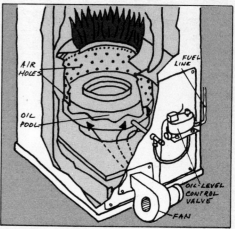
Pot-type oil burner

To save energy and cut fuel costs

- Have your home *fully* insulated.
- Install a humidifier.
- Install well-constructed, tight-fitting storm doors and windows. Make sure regular windows are tightly closed.
- Add weather stripping to doors.
- Check and, if necessary, add caulking around doors and windows.
- Set thermostats back 5 to 8 degrees in the evening or whenever the house will be unoccupied for more than a couple of hours.
- If you have a fireplace, make sure the damper is fully closed.
- Make sure your heating system is clean, properly adjusted, and well-maintained.

Oil burners

Oil burners have been widely used in home-heating systems for more than fifty years. Oil-fired heat producers work well with either hot-water, steam, or hot-air systems.

There are two types of oil burners in general use: the pressure type and the pot type. Both of these burners are located outside of the furnace or boiler fire box and so are relatively easy to service.

A third type of burner—called a rotary burner—has just about disappeared from use for home heating but is still used in some large building installations. The entire rotary burner unit is located within the fire box. This makes even routine service and adjustment rather complicated.

Pressure-type burners: Pressure-type burners pump oil, under pressure, through a small nozzle to form an oil spray. Air is mixed with the oil spray to form a combustible mixture. High voltage is applied to two electrodes and a spark jumps between them to ignite the mixture.

When operating properly, the oil and air mixture will continue to burn until the burner is turned off. Vanes mounted at the tip of the burner deflect the flame into a swirling form which allows more complete combustion of the mixture. As the combustion process becomes more complete, more heat is extracted from the fuel and less soot remains.

The operating parts of a pressure burner are an electric motor and a transformer. The motor drives a pump which creates the oil pressure and a fan which draws air into the burner. The transformer steps up the 120-volt house power to a high enough voltage to ignite the flame. Pressure-type burners can be turned on and off as needed, by means of a manual switch.

Pot-type burners: In this type of burner, a quantity of oil flows into a container (pot) in the fire box. The oil is ignited by a pilot flame. Once ignited, the heat causes the oil to vaporize. Air from a small blower mixes with the oil vapor and causes it to ignite. Heat level is controlled by regulating the flow of oil to the pot. When the oil flow is changed, the air flow must also be changed to maintain good combustion.

Pot-type burners are simple in operation and easy to maintain. Initial adjustment, however, is tricky and should be left to an experienced serviceman.

Stack relay

In either type of burner, if ignition fails and oil continues to be pumped into the fire box, the fire box and surrounding area are soon flooded. To prevent this, all burners have a device to turn off the oil flow if ignition fails. The most common device is the stack relay, which consists of a heat-sensitive switch, a relay circuit, and a restart button. The relay circuit and the restart button are enclosed in a metal box.

The stack relay is mounted on the sheet-metal stack that runs from the top of the furnace to the chimney flue. A hole is cut in the stack so that the heat-sensitive switch can project into the stack. When the burner is on and working properly, heat from the fire box keeps the heat-sensitive switch closed. This, in turn, keeps the relay circuit energized and maintains electric power to burner motor. If the flame goes out, the heat-sensitive switch opens and shuts off power to the burner fan and oil pump.

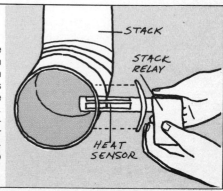

How to maintain an oil burner

Oil burners should be adjusted, inspected, and cleaned by burner-service personnel at the beginning of each heating season. Once adjusted, there are a number of things the homeowner can do to maintain efficiency and correct minor defects.

The following items should be serviced once about the middle of each heating season. If you have a burner-service book, read and follow the detailed instructions given for each of these items. If you need additional help, ask your serviceman to show you what needs to be done.

Turn off the burner master switch and clean the blower fan blades. The blades can be reached through the air-intake opening next to the fan. If you must remove a plate partially covering the air-intake opening, mark the exact position of the plate beforehand so it can be correctly replaced. Brush dust and dirt from the fan blades.

Turn off the burner master switch and take out the stack relay by removing the two mounting screws. Clean the heat-sensitive switch element. Use a small brush dipped in detergent and warm water. Hold the stack relay with the heat-sensitive switch pointed down so water does not run into the relay box. Shake off excess water before replacing the stack relay.

With the burner turned on and operating, check the position of the draft vane (see box, below).

Once or twice each heating season, step outside and take a look at the chimney. Do this at a time when you know the burner is on. If smoke or soot can be seen coming out of the chimney, the oil-spray-and-air mixture is not burning properly. Notify your burner service company.

How to restart an oil burner

If your burner does not start automatically as it should, make a few quick checks before you call your service company.

Check that the burner master switch has not accidently been turned off.

Check the circuit breaker or fuse that protects the oil burner circuit. If the circuit breaker is tripped to OFF or the fuse is blown, you may have an electrical short in the burner circuit. Notify your burner service company. Do not reset the circuit breaker or replace the fuse until the cause of the turn-off is known.

Check the oil level in your tank. If the tank is empty, the burner should not operate. If the float gauge indicates oil, check it by tapping lightly to be sure it is not sticking.

Check the thermostat setting against the indicated temperature. If the setting is five degrees or more higher than the indicated temperature, the burner should start. Remove the thermostat cover and check whether inside electrical contacts are exposed or sealed in glass. If the contacts are exposed, you can clean them by rubbing a crisp dollar bill between them. Do not use a file or any abrasive to clean contacts. If contacts are sealed in glass, no maintenance can be done. If the thermostat is defective, it must be replaced.

If the thermostat seems okay, if there is oil in the tank, and if the circuit breaker or fuse is okay, press the restart button on the stack relay once. If the burner starts, fine. For the next hour or so, check the burner every 10 to 15 minutes. If you see any evidence of oil leakage or if the smell of oil is unusually strong, turn off the master switch and notify your service company.

If pressing the restart button does not start the burner, the ignition system may be at fault. Notify your burner service company.

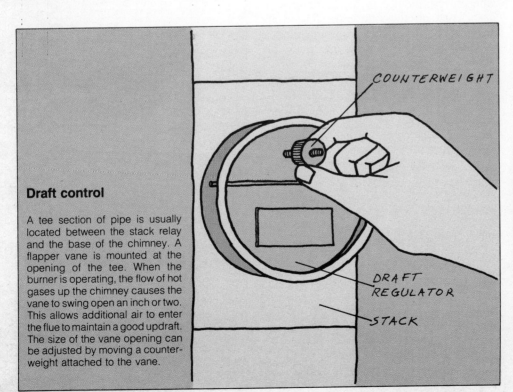

Draft control

A tee section of pipe is usually located between the stack relay and the base of the chimney. A flapper vane is mounted at the opening of the tee. When the burner is operating, the flow of hot gases up the chimney causes the vane to swing open an inch or two. This allows additional air to enter the flue to maintain a good updraft. The size of the vane opening can be adjusted by moving a counterweight attached to the vane.

Add a few drops of light oil to each burner motor-bearing cup. A few drops are all that should be added, since over-oiling can lead to serious motor trouble.

Gas burners

What it takes

Approximate time: One hour to clean pilot jet and relight and adjust pilot flame.

Tools and materials: Piece of soft copper wire, matches.

1. Put thermostat on lowest setting, usually marked OFF.

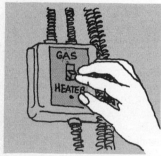
2. Switch off current to furnace.

3. Shut off gas to furnace but not to pilot light.

4. Light pilot, start current, open main valve, set thermostat.

A gas-fired heat producer operates much like a gas range. The gas-fired furnace is larger and, because it is out of sight, contains safety features not required on cooking ranges.

The gas is ignited by a pilot light. An automatic system turns off the gas supply if the pilot light goes out.

A gas furnace or boiler may be used with an air, water, or steam-heating system. Gas heating has the advantages of clean, low-pollution burning and no home storage of fuel. In cities served by gas utility companies, mains carry gas to the home. Where bottled gas is used, containers of gas are delivered by truck. In most parts of the country, gas is more expensive than oil or coal.

Gas-fired furnaces require less service than other types of heat producers. It is customary in most localities for servicing to be done by the utility company which supplies the gas. The customer is charged for replacement of defective or broken parts, but not for routine service.

The home owner's maintenance of the gas furnace itself consists of relighting and adjusting the pilot light and cleaning the pilot jet.

Gas-fired furnaces are vented to the outside through stack connections from furnace to chimney, as are other furnaces. The only difference is the addition of a draft hood on the vent pipe. The draft hood is designed to prevent chimney downdrafts from blowing out the pilot light. The draft hood is effective most of the time, but strong, sudden downdrafts can still occasionally blow out the pilot.

A device called a thermocouple is located near the pilot flame. The thermocouple generates a small amount of electric current when heated by the pilot flame. This current is used to energize a solenoid valve that keeps the main gas supply line open. If the pilot flame goes out, the thermocouple cools and no longer generates electric current. The solenoid valve closes and turns off gas to the main and pilot jets.

The detailed procedure for cleaning the pilot jet and relighting and adjusting the pilot flame varies for each make and model of gas burner. The step-by-step procedure, together with the names and locations of controls, is given on a metal plate attached to the burner unit. The general procedure for all units is as follows:

1. Set the room thermostat to the lowest temperature, or to the OFF setting if it has one marked on it.
2. Turn off electric power to the furnace. This can be done by turning off a master switch or by setting a circuit breaker to OFF or removing a fuse.
3. Set the gas supply valve to the position that turns off both the main gas supply *and* the pilot supply.
4. Pull off the metal deflector attached to the top of the pilot jet.
5. Clean the jet by inserting a piece of soft copper wire in the pilot hole. Replace the deflector.
6. Set the gas supply valve to the position that keeps the main gas supply shut off, but turns on gas to the pilot jet.
7. Light the pilot. Adjust the pilot control valve for a flame about 1¼ inches high.
8. Set the gas valve to the full open position and turn on electric power.
9. Set the thermostat to the normal house temperature.

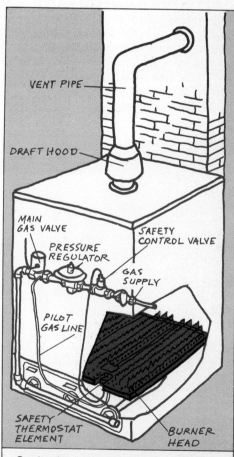
Gas is mixed with air and fed to a burner in the furnace. The burner may have a single large flame, a pattern of small jets, or a ribbon flame.

78 HEATING

Electric heating

Electric heating is easy to control, requires little or no maintenance, and is clean and quiet. The only disadvantage of electric heating is the high cost in most areas. To be economically practical, electric heating requires an exceptionally well-insulated house. It is best used in areas having moderate winter temperatures.

Electricity can be used as the heat producer in hot-water or hot-air systems. Conventional hot-water baseboard systems using gas or oil can be readily converted to electric heating. No change is required in the system of pipes and convectors. The oil or gas boiler is replaced by an electric water-heating unit. The electric unit requires much less space and needs no flue connection. The unit can be mounted on a wall or installed in an attic or basement area.

Electric hot-water systems are controlled by a thermostat and an aquastat (see page 80) in the same manner as an oil or gas-fired system. Control of the individual convectors will depend upon the arrangement of the circulating loop (see page 80 for more on circulating loops).

Similarly, the hot-air and return-air ducts of a conventional hot-air system can be used with an electric heat producer. The circulating fan draws in cool air and passes it through a network of resistance wiring. When power is on, the wiring is hot and it heats the air as it passes through. The heated air is then carried by ducts to the living areas of the house, as it is in an oil- or gas-fired system.

Self-contained electric baseboard-heating units can be used for special areas or as auxiliary heaters. In a mild climate, these units can be used as the main heat source. Several types are available.

One type of baseboard unit uses a single heating element much like an electric furnace. The heating element runs through a ceramic-lined metal tube. The tube is surrounded by metal fins The heating unit is enclosed in a metal housing designed to resemble conventional baseboard. Openings at the bottom and top allow air to flow through to heat the room. Units are made to operate on either 120- or 240-volt power. The 240-volt unit, of course, produces more heat. Each unit has its own switch, thermostat, and over-temperature control to prevent burn-out if air flow is blocked.

Another baseboard unit, similar in appearance to the heating-element type, contains a liquid in a sealed tube. The sealed tube is surrounded by fins. The liquid is electrically heated. The fluid, in turn, heats the fins and the air that passes through. Because the fluid retains heat for a time after power is turned off, this unit offers a somewhat more even heat flow.

In areas where extra heat is required for short periods of time (bathrooms, for example), self-contained wall units can be installed. These units use an electric heating element similar to the baseboard unit. In place of fins, the heating element is formed into a grid. A fan is used to move air through the unit. The units are designed to fit between wall studs. They, too, are equipped with manual on-off and temperature controls.

Electric radiant heating uses resistance wiring to produce heat. Resistance wire heats up when current passes through it. In principle, it is the same as the heating element in your electric toaster.

The heating wires are embedded in some insulating material—such as plaster—in walls or ceilings. When power is turned on, the wire heats up and radiates heat to the people and objects in the room.

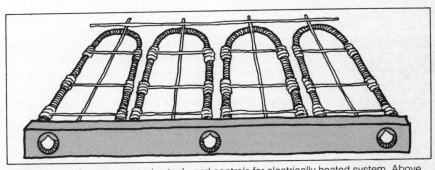

At left, heat exchanger, expansion tank, and controls for electrically heated system. Above, radiant heating elements.

Electric baseboard heater

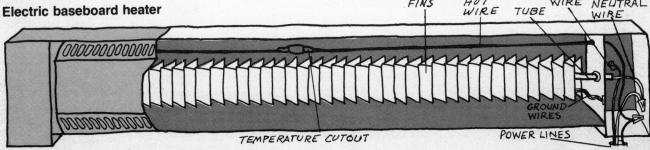

Heat exchangers, distributors, and conduits

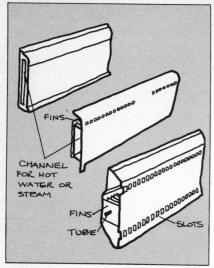

Baseboard convectors are generally used in series loops. The convectors transfer heat from the water circulating through them to the air in the room.

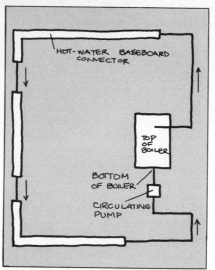

In a series loop, each baseboard unit is part of the loop. There is no way to turn off individual convectors. Thus, heat control is limited in this system.

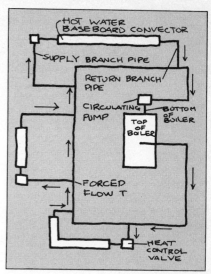

In the two-pipe system, one pipe supplies hot water to convectors and one returns cooled water to the boiler.

Forced hot-water systems. These systems work by heating water as high as 240 degrees and circulating this water through pipes, radiators, and convectors.

Any heat producer—oil, gas, coal, or electricity—can be used with a hot-water system. Each is economical, efficient, and heats domestic hot water in the same boiler. However, hot water has the disadvantage of higher initial cost than hot air, and the difficulties associated with a piped-water system.

Hot-water heating is a closed-loop system. The loop—consisting of heat exchanger, piping, and radiators—is filled with water. Water is pumped through the loop by a circulator pump. An expansion tank is included to allow for expansion of water and to maintain it under pressure.

The tank contains trapped air and water. As heated water expands, it compresses the air. Air resists compression and maintains the water under pressure. Because water in the system is under pressure, it can be heated above the normal boiling point and not become steam.

The heat producer is located in the base of the hot-water boiler. Heat and gases flow over closed sections in the heat exchanger. Water for the heating system flows through the closed sections and absorbs heat. The heat exchanger also contains coils through which cold water flows to be heated for household use.

It is important to remember that domestic hot water and heating hot water are independent systems, sharing only boiler heat.

The simplest system for distributing hot-water heat is the series loop. However, heat control is limited in this system, since there is no way of turning off individual baseboard convectors. The two-pipe system allows heat to be controlled as desired, because one convector can be adjusted without affecting the others.

A variation of the series-loop and two-pipe systems uses convectors in parallel with the circulating pipe. When a convector is turned off, hot water is diverted to the main line and continues to the next one.

Zoned heating. This is a combination of two systems operating from one boiler. The systems may be independent, each having a circulator, thermostat, and flow controls. A variation uses one circulator for both zones, created by dividing the main supply into two loops. The two loops are joined back to the boiler when they have completed their run. Flow is controlled by electric zone valves with separate thermostats.

Hot-water-system gauge. A pressure and temperature gauge is mounted on the side of the boiler. High and low heating-system water-temperature readings should be within 5 or 10 degrees Fahrenheit of the aquastat settings.

The altitude scale is a reference scale and has two pointers. One is set during installation or major servicing, and, once set, does not move during operation of the system. The other, movable pointer indicates water pressure. The correct readings and limits for this scale vary with sizes and types of systems. When your system is serviced, determine your normal readings. Then make a sketch of the gauge on a small card and mount it nearby for future reference.

An aquastat senses water temperature in the boiler and turns the heat producer on when the temperature falls below a preset limit and off when it reaches a preset high. Two types of aquastats are in general use. The newer model has two dials under a removable front cover, one setting the high limit, one the low. Dials can be adjusted to suit heating requirements. A high setting of 160 degrees and a low of 120 degrees provide adequate heat in moderate weather; for colder weather a high of 180 degrees and a low of 150 degrees provide more heat. These levels are not the same for all systems. Your service company can advise on correct limits. Older aquastats have one high setting indicated by a dial and pointer. The low limit is preset.

Routine maintenance

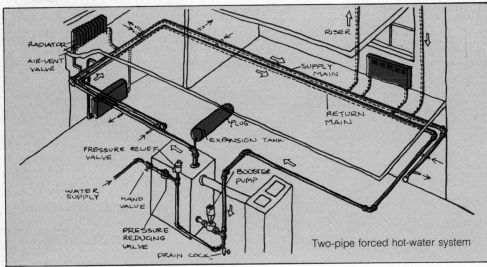

Two-pipe forced hot-water system

Bleeding the line
Some or all of the radiators or convectors in a hot-water system have small valves at one end. These allow trapped air to be released. Have a cup handy. Use a screwdriver (or vent key) to open the valve. Escaping air will cause a hissing sound. Water will spurt out when air has been released. Catch the water in the cup. Turn off the valve when a solid stream of water is coming out.

Circulator lubrication
At the beginning of each heating season, add a few drops of oil to the oil cups at each end of the motor. SAE grade 20 oil is usually specified somewhere on the motor housing. The circulator may also contain an oil cup. If so, add a few drops of SAE grade 20 oil to it as well.

Draining the system

Any time there is danger of freezing temperatures within the house—when electric power fails, for example—the system should be drained to keep pipes from bursting.
1. Turn off the heating system by turning off the master switch. Also set the heating-system circuit breaker to OFF or remove the heating system fuse. This should be done, even though electric power is off. The heating boiler could be seriously damaged if power was restored while the system was drained.
2. Allow the water in the system to cool down to a lukewarm or cool temperature.
3. Turn off the valve from the household's cold-water supply.
4. If practical, attach a hose to the drain at the lowest level in the system. This drain is located below the circulator, near the base of the boiler.
5. Run the other end of the hose to a basement floor drain. If no floor drain is available, the system can be drained into a pail. You will have to empty the pail several times.
6. Open the drain and open the air vents on the highest radiators.
7. If the system is being drained because of the danger of freezing, leave all drains open.

Flushing and refilling

1. With a hose attached to the boiler drain (or a pail beneath it), turn on the valve from the household's cold-water supply.
2. This will flush clean water through the boiler. Continue flushing until the water coming out of the boiler is clean.
3. To fill the system, close the boiler drain valve. Also, close the air vents you opened to drain the system.
4. Restore power to the system. Adjust the thermostat to start the burner. (You may also have to press the restart button on the stack relay.)
5. After the system has operated for several hours, bleed the radiators or convectors to free trapped air.

Draining the expansion tank

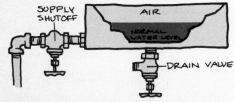

The expansion tank should be about half full of water for proper system operation. Over a period of time, the water level tends to rise because the system water absorbs small quantities of air. When too much air is absorbed, the water pressure may become high enough to open the pressure-relief valve, allowing hot water to flow out of the valve. If this should happen, turn off the heat producer immediately. Allow the system to cool. The pressure-relief valve will close automatically when the pressure drops.

The following procedures will prevent pressure build-up:
1. Attach a hose to the drain valve at the bottom of the expansion tank. Run the hose to a pail or drain.
2. Turn off the expansion-tank inlet valve.
3. Open the drain valve and allow the tank to drain completely. Disconnect the hose and close the drain valve. Open the tank inlet valve. Water will flow into the tank and compress the air in the tank to establish normal system pressure. The pressure reading on the boiler gauge should agree with the normal reading marked on your card.

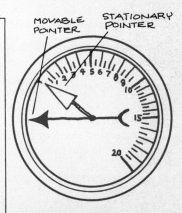

Boiler gauge showing pressure reading. If movable pointer drops below stationary pointer, as it has here, system needs more water.

HEATING 81

Forced hot-air systems

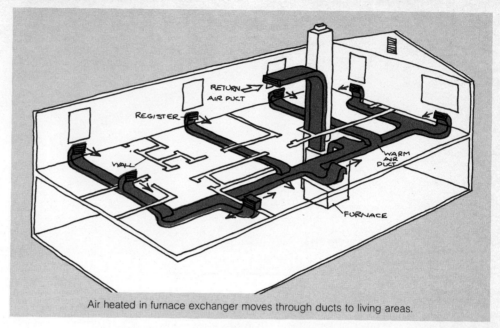

Air heated in furnace exchanger moves through ducts to living areas.

What it takes

Approximate time: Two hours.

Tools and materials: Flashlight or utility light, mirror, screwdriver, T-square or carpenter's level.

One of the simplest and most widely used heating systems in the United States is forced hot air. Forced air systems usually have oil or gas burners as the heat producers. A squirrel cage or centrifugal fan, driven by an electric motor, is the heat distributor.

In a typical basement hot-air furnace, heat from the burner passes through metal pipes in a heat-exchanger section. Hot gases from the burner heat the pipes to temperatures of several thousand degrees. The hot gases are ducted to the flue and vented to outside the house. Air is heated as it passes over the hot pipes of the heat exchanger.

The heated air is then carried by ducts to the living areas of the house. In some systems, the heated air first goes to a large sheet metal enclosure called a *plenum*. Ducts to various parts of the house are connected to the plenum.

When the furnace is on the same level as the living area—as in homes built on a poured concrete slab—the air flows through the furnace from top to bottom. The fan draws air in at the top. The air is forced through the heat exchanger and then into the ducts that are embedded in the concrete slab. Forced-air systems are controlled by a room thermostat. When heat is called for, the thermostat completes an electrical circuit to turn on the oil or gas burner.

An additional device is needed in hot air systems to control the blower. If the blower were to start immediately before the heat exchanger reached operating temperature, cool air would be blown into the living areas. A device called a fan-and-limit switch senses the temperature in the heat exchanger and turns on the blower only after the exchanger reaches a preset temperature. The fan-and-limit switch also turns off the burner when the temperature in the heat exchanger reaches the high limit. The switch allows the blower to continue operating as long as heat remains in the exchanger.

Maintenance and adjustment

The most common problem with forced hot-air systems is insufficient or uneven heat. In correcting this condition, see if air flow is blocked. If not, adjust flow controls to get the desired heat, as follows:

1. Remove and clean or replace the air filter at the furnace air inlet. A dirty filter restricts air flow and causes both a reduced air volume and slower flow.

2. Check for blockage in ducts. This can be caused by damage or by an accumulation of dirt and lint. Remove registers. Have someone hold a light at one register. Aim the light toward the next register on that duct line. At the next register, look back toward the light. This will show up significant blockage, even with a bend in the duct.

3. Damper function. Heat flow can be adjusted if the filter is clean and the ducts are clear. Dampers are metal vanes in the air ducts. A control handle on the outside of the duct allows you to adjust for more or less air through the duct. The angle of the control handle is the same as that for the vane. When the control handle is horizontal, the damper is fully open; when vertical, the duct is closed.

4. Adjusting dampers to suit your individual needs is a matter of trial and error. Two guidelines may be helpful. First, reducing air flow in one duct increases the air flow in another duct, if the same volume of air is coming into the system. Second, dampers are installed in ducts near the furnace or main plenum. Reducing the air flow in a duct reduces the heat coming from registers farthest from the furnace more than from those nearest.

Adjusting blower speed

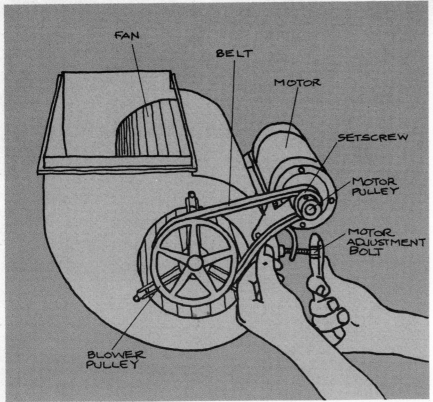

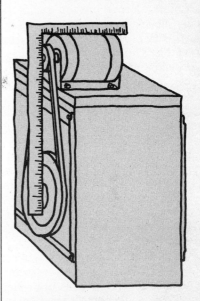

Belt tension. If blower speed is too low, air flow is reduced and rooms will not heat as fully or as rapidly as they should. If blower speed is too high, the blower will be noisy. The high speed will not improve heating.

Blowers are driven by electric motors. A small pulley is attached to the motor shaft and a larger pulley is on the fan shaft. The fan is driven by a belt between the pulleys. If the belt is too loose, it will slip and fan speed will be reduced. With the fan turned off, midway between the pulleys you should be able to depress the belt no more than ¾ inch to one inch. More slack than this means slippage is likely. Less slack usually means noise and rapid wear.

Belt tension is adjusted by turning an adjustment bolt located on the motor-mounting frame, which moves the motor toward or away from the fan.

Pulley alignment

For the least noise and bearing wear, the motor and fan pulley should be aligned. Use a straight-edge, T-square, or carpenter's level to check alignment. To adjust, loosen the set-screw on the inner section of the motor pulley. Move the pulley in or out as necessary. Tighten the set-screw.

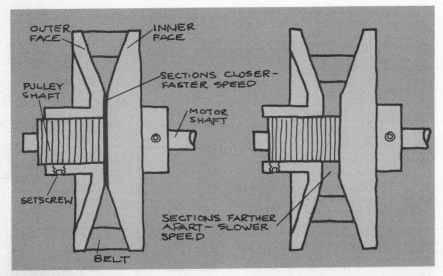

Blower speed. On some blower units, the motor pulley may be adjusted to change the fan speed. The pulley consists of two sections mounted on a threaded stud. Both sections have collars with setscrews. When the setscrew in the outer collar is loosened, the outer section can be turned while the inner section is held stationary. In this way, the space between the sections can be changed. When the sections are brought closer together, the belt rides higher on the pulley. This increases the speed of the fan. If the space between the sections is made larger, belt rides lower; speed is decreased.

Steam-heating systems

Steam-heating systems, once widely used, have largely been replaced in recent years by less costly hot-air or hot-water systems. However, steam heat is durable and efficient.

The system works by heating water in a boiler until steam forms. The steam, being lighter than air, rises naturally through the pipes to the radiators. Vents on the radiators allow air to escape as the steam rises. The steam gives up heat to the radiators. The radiators, in turn, heat the surrounding area. As the steam cools, it condenses back into water. The water flows back through the same pipes to the boiler where it is reheated, and the cycle is repeated.

The entire steam system—boiler, piping, and radiators—must be solidly constructed to withstand steam pressure and temperature changes. In addition, for maximum efficiency, steam pipes must be insulated to prevent excessive heat loss before the steam reaches the radiators.

To monitor and control the system, the steam boiler contains a water-level gauge and a pressure-relief valve. The water-level gauge is a glass tube containing water. The level in the tube represents the level in the boiler. When the water in the gauge is about midway between the top and bottom, the correct amount of water is in the boiler.

The pressure gauge registers the steam pressure within the boiler. When the system is operating properly, the gauge should indicate 10 to 12 psi (pounds per square inch). If the pressure exceeds a predetermined level, the relief valve opens to allow steam to escape. When the pressure falls to a safe level, the relief valve closes automatically. Many steam boilers also contain a float valve to cut off the system if the water level in the boiler is too low.

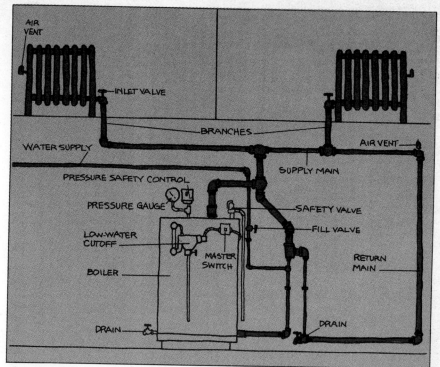

Radiators

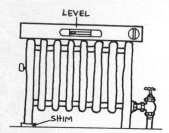

Radiators can be checked with a carpenter's level. The radiator should slope toward the shutoff valve and pipe end. It can be shimmed with pieces of wood or metal to get the proper slope. Also, make sure the shutoff valve is either fully open or fully closed.

Hammering can also be caused by sagging pipes. The noise will localize the trouble to some extent. Check the piping in the vicinity of the noise. Pipes should slope toward the boiler. Sagging can generally be corrected by adding additional pipe-strap hangers.

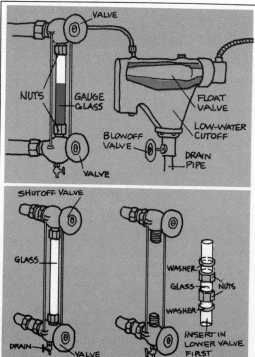

Routine maintenance

At least once a week, check the water-level gauge. If the water is below the midpoint of the tube, open the boiler-inlet valve and fill to the proper level. While some boilers have an automatic fill valve, a weekly check is still advisable to guard against failure of the automatic valve. If, at any time, no water is visible in the gauge, turn off the heat producer *immediately*. Allow the boiler to cool before adding water.

After a time, the water gauge may become so dirty that it is difficult to read. When this happens, turn off the valves at the top and bottom of the gauge and loosen the collar nuts. You can then lift the glass tube out. Use a bottle brush to clean sediment out of the tube. Replace the tube, tighten the collar nuts, and open the valves.

Sediment will also collect in the float valve, which can be cleaned by flushing. Open the blow-off valve and allow water to flow out until no sediment is visible.

The most common problem with steam systems is noise (hammering). The noise is usually caused by poor drainage which allows water to collect in radiators or pipes.

Fireplaces

In modern homes, fireplaces are generally more decorative than functional. However, as fuel costs become an increasingly greater economic burden, all forms of auxiliary heating must be considered.

Most building codes require that masonry fireplaces be completely self-supporting. The fireplace must have its own masonry footing. This means that adding a fireplace to a home that does not have one represents a major expenditure.

If a fireplace is to be more than decorative, it must be well designed. There is a definite relationship between the front opening and the flue opening. The former should be about ten times larger than the latter. If the cross-sectional area of the flue opening is 100 square inches (10 by 10 inches, for example), the front opening should be approximately 1000 square inches and should be wider than it is high (10 inches wide by 25 inches high, for example).

For safety and good operation, fireplaces should have a damper and a smoke shelf. When the fireplace is in use, the damper can be adjusted to control the updraft and so control fire. When the fireplace is not in use, the damper should be fully closed to prevent loss of room heat through the flue. The smoke shelf is a projection running horizontally across the smoke chamber. It deflects downdrafts to prevent smoke and sparks from being blown into the room.

Cleaning chimneys. The first step in chimney cleaning is to seal the fireplace opening with paper and tape to prevent dirt from getting into the room. Leave the damper open. Fill a burlap bag with old rags and a couple of bricks so that it blocks the chimney opening. Tie a rope on the bag and lower it down the chimney from the roof. Pull the bag up and lower it again several times to loosen soot and dirt on the sides of the flue.

Allow time for dust and dirt to settle, then remove the paper from the opening. Brush dirt and soot off the smoke shelf and damper. Brush all dirt into the ash pit opening. Empty the ash pit.

Soot and smoke stains around the fireplace opening can generally be removed with a strong solution of detergent and hot water.

Tip on lighting a fire. In cold weather, the flue is filled with heavy, cold air. Unless quickly heated, this cold air prevents a natural updraft from starting and causes the fire to die out.

A large, hot, initial flame will heat enough air to push out the cold-air column and get a natural draft going. Use plenty of light kindling or several wads of newspaper to get a large, hot flame going quickly.

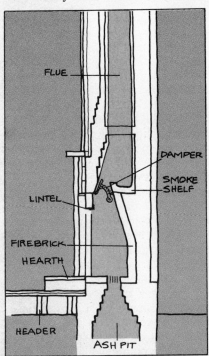

A ground-floor fireplace should have a damper and a smoke shelf for best operation. An ash pit, with a clean-out in the basement, is a great convenience but not essential.

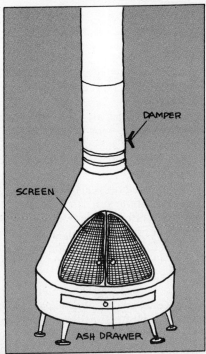

Freestanding fireplaces are essentially small, metal stoves. They are a less costly substitute for masonry construction and much easier to install. They are adequate for supplementary heat, hence suitable for vacation homes.

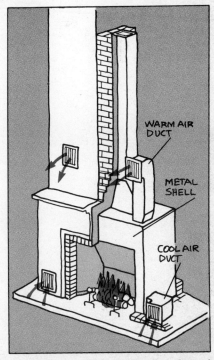

Fireplaces can be built around preformed double-walled sheet-metal forms which establish proper proportions for the fireplace and improve heating efficiency by providing for air flow through the double-walled area.

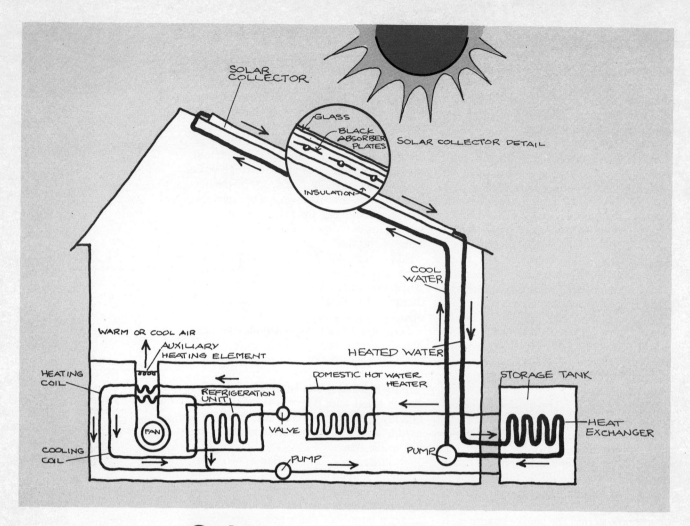

Solar heating

Using solar energy for everyday heating has been talked of for many years. It is now—at least partially—a reality.

More than 200 companies are currently marketing solar-heating systems of various capacities, starting with those designed only to heat some of the water required for domestic use. Cost savings from solar heating vary according to geographical location. The more days of sunshine you can expect in a year, the more effective your solar-heating unit will be. Many solar-heating units are sold in kit form for installation by the home owner. The installation is fairly easy, and the savings are substantial.

The most widely used solar water-heating system operates as a simple preheating unit. A liquid is pumped to the roof, heated by the sun, and returned to a heat exchanger. In the heat exchanger, the heated liquid transfers heat to the domestic water supply. Water is then piped to a conventional water heater. The water heater raises the temperature of the water as necessary and stores it for use as needed.

The cost saving results from the preheating of the water before it reaches the water heater. The water heater has to raise the temperature of the water only a few degrees when the solar unit is working at peak efficiency. The conventional water heater also maintains the water temperature during evening hours or periods of little sun.

The unique element in the solar-heating system is the collector panel installed on the roof. A typical collector panel measures about eleven feet long by three feet wide. The collector panel consists of a heat-absorbing surface covering a continuous length of coiled tubing. The heat-absorbing surface is protected from the weather by a curved plastic shield. The shield also creates a "greenhouse" effect that retains heat in the collector. Each collector panel weighs about 75 pounds. The number of collectors used in a system varies according to the space available and the heating capacity desired. Two collectors of the size indicated above are recommended for an average house.

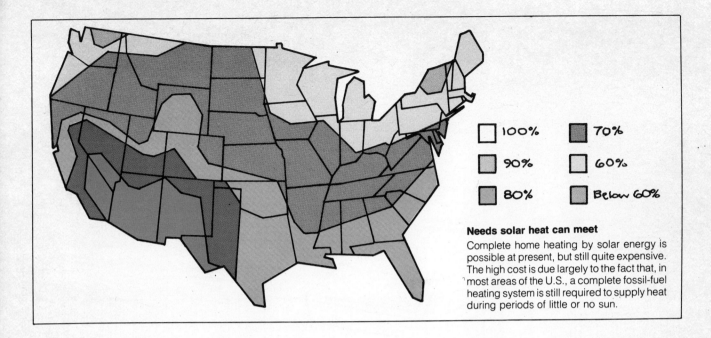

Needs solar heat can meet
Complete home heating by solar energy is possible at present, but still quite expensive. The high cost is due largely to the fact that, in most areas of the U.S., a complete fossil-fuel heating system is still required to supply heat during periods of little or no sun.

The collectors should face south and should be mounted at an angle of 30 to 50 degrees above horizontal. If a flat roof is used or a roof with a shallow pitch, the collectors can be mounted on wooden frames to get the best angle. It is not essential that the collectors face directly south. However, deviations of more than 30 degrees from due south will cause a significant reduction in efficiency.

In addition to the two collector panels, a typical solar water-heating kit includes a storage tank with heat exchanger, circulating pump, thermostatic control, expansion tank, check valve, and antifreeze solution. The antifreeze circulates in the collector and heat-exchanger line. The use of antifreeze solution protects the system against unseasonable temperature drops. Kits do not normally include the piping to run from the collectors to the heat exchanger, or lumber for mounting the collectors on frames attached to the roof.

Solar-heating units designed to heat swimming-pool water are also available. These units are similar to domestic water-heating units. The principal difference is that swimming-pool water flows through the collectors and is returned to the pool. Heat exchangers and storage tanks are not required. However, all collector tubing and system piping must be fabricated of material that can withstand the corrosive effects of chlorinated water. Collectors for domestic hot water and swimming-pool use cannot be interchanged.

Schematic of typical water-heater installation for solar collector

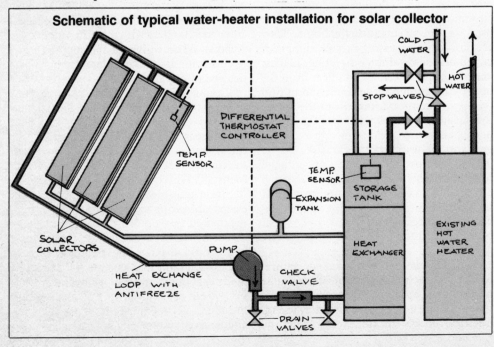

Heat pumps

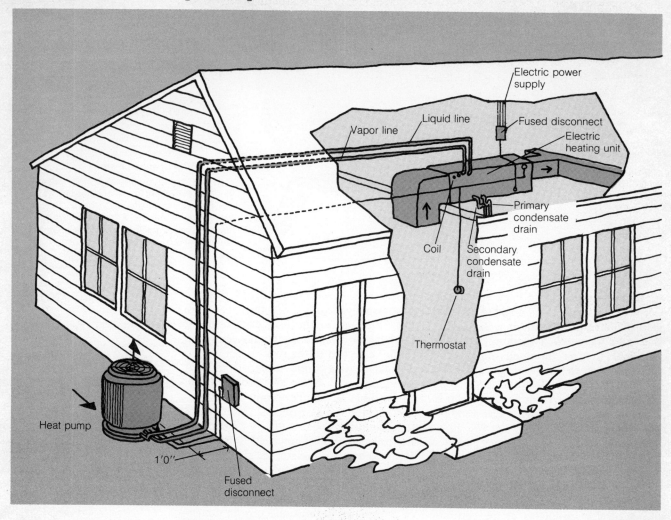

Heat from the outside air

The drawing above shows typical piping and wiring for a heat-pump system installed with a supplementary electric heating system. Most heat pumps of this type have the coil unit inside and the heat exchanger outside the house. Some pumps, used extensively in mobil homes, have both contained in one unit.

Heat pumps, originally developed from air-conditioning technology, were first used for cooling air inside the home. They are now used for heating as well, since they can automatically change direction. During hot weather they can extract heat from the inside air and pump it outside; and in cold weather, take the warmth from the sun in the outside air and pump it into the house. Some heat pumps are designed to exchange heat with natural bodies of water.

Heat pumps differ from conventional heating systems because they use existing heat; moving some of the natural warmth of the outside air into the house. More conventional systems produce their own heat by burning oil or gas, or by using electricity to heat wires.

Although there is always *some* natural heat from the sun in the outside air most heat pumps operate efficiently (that is, produce more heat energy than they use up electrical energy to run) only when the outside temperature is over 30 degrees Fahrenheit. For this reason, heat pumps are used in tandem with other home heating systems in most parts of the country.

Heat pumps are considerably more expensive to buy and install than traditional room air conditioners. Like central air conditioning, they are best installed as part of the basic temperature control system when the house is built. Their plumbing, electrical, carpentry and duct systems make them too complex for do-it-yourself homeowners to install. The components of a representative heat-pump system, coupled with an electrical-heating system, are shown in the diagram at the top of this page. To determine the practicality of using a heat pump, you should balance the purchase prices, installation costs and operational expenses of the heat pump and a supplementary heating system against the cost of purchasing, installing and operating alternative heating and cooling systems.

When are heat pumps most efficient?

The map below shows the approximate percentage of time (expressed in decimal portions of a year) that a heat pump, piggy-backed on a conventional heating system, will be in operation to heat or cool your home to keep it at a comfortable temperature. The percentage of time the heat pump is in operation depends on the percentage of time the outdoor temperature is above 30 degrees Fahrenheit. When the temperature drops below that, the heat pump shuts off automatically and signals the conventional heating system to begin normal operation.

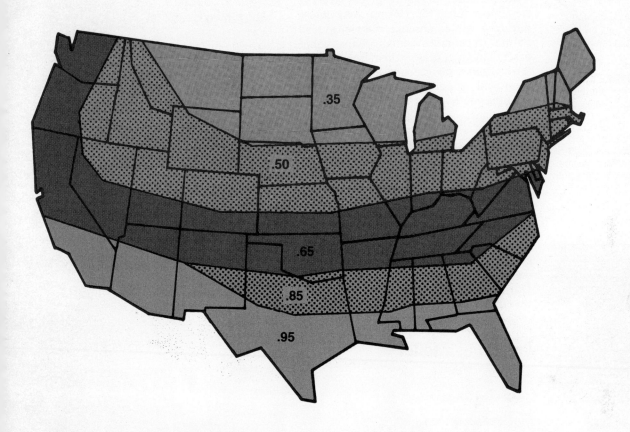

How a heat pump works

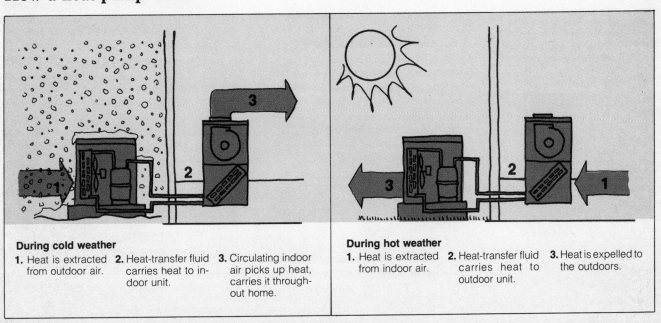

During cold weather
1. Heat is extracted from outdoor air.
2. Heat-transfer fluid carries heat to indoor unit.
3. Circulating indoor air picks up heat, carries it throughout home.

During hot weather
1. Heat is extracted from indoor air.
2. Heat-transfer fluid carries heat to outdoor unit.
3. Heat is expelled to the outdoors.

HEATING

10. PROJECTS

Cold-weather house shutdown

The checklist at right gives step-by-step procedures for closing your house during a winter absence. Obviously, all fixture traps must contain some liquid other than water to keep them from venting sewer gases into the house while you're away, and the liquid must not evaporate in the time you're gone. Consider the following options:

Remove all trap water by draining or blowing out sink and lavatory traps. Refill all with kerosene (hard on rubber trap gaskets) or glycerine (more costly). Here's an alternative: fill traps with a mixture of a little kerosene and a lot of denatured alcohol. The kerosene will keep the alcohol from evaporating. Another alternative is to leave water in the traps and pour a cupful of auto antifreeze in each. Also, pour at least one quart in each toilet and basement drain. During reopening, be sure to fill the water-heater tank and hydronic heating-system boiler before you fire them up again.

What it takes

Approximate time: Complete house shutdown for a wintertime absence may take about two hours.

Tools and materials: You'll need a pair of pliers, some trap antifreeze (see text), a length of garden hose, and perhaps pipe tools.

Planning hints: Be sure that you are through using water before you start the shutdown. Posting a checklist of what was done may be a good idea, so that you can quickly get the house reopened when you return.

Shutdown checklist

1. Turn off house water supply at underground street valve.
2. Open all faucets and outdoor hose spigots to drain.
3. Turn off the fuel to the water heater, connect a garden hose to its drain and drain the tank.
4. Flush toilet and sponge out remaining water from tank.
5. After all draining ceases, open stop-and-waste valve(s) in crawl space or basement.
6. Drain or blow water out of fixture traps, including toilet.
7. If there are any low spots in horizontal piping, take joints apart and drain by hand (or see Practical Pete's idea at left).
8. Drain hydronic heating system, including boiler. Turn off fuel first. Be sure that the boiler's water-supply valve is open to drain its piping. Open radiator air valves.
9. Drain appliances: washer, dishwasher, furnace humidifier.

Getting out from under

Let me clue you in on a fast way to clear your house water pipes of water. A plumber who owns a vacation cabin in Michigan showed it to me. After the water heater has been drained in the normal way and everything else is empty, there may be water left in some of the low spots in horizontal mains. To get it out fast, just stick the end of a running air-compressor hose into an opened outdoor hose faucet. The air pressure will shoot the water out wherever there is an opening. When only air comes out, you're done. This method sure beats crawling under the house to drain your pipes. Neat trick, I've found!

Practical Pete

Preparing a toilet

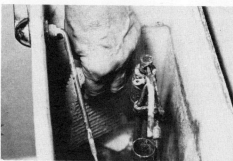

1. Don't leave any water in the toilet tank that might cause it to freeze and crack. Flushing will remove most water. Then sop out the rest with sponge.

2. Pour permanent-type auto antifreeze coolant into the toilet bowl to protect it. One-third gallon—a 33% mixture—should protect the bowl down to about zero degrees Fahrenheit.

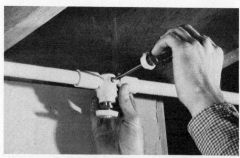

3. Drain all house piping in crawl space, basement, and attic. Low point of water-supply piping system should have a stop-and-waste valve where it can be drained. This one contains a drain screw.

4. Finally, pour about 4 ounces of auto antifreeze coolant into each fixture trap. Remember that a 1:2 antifreeze-water mixture protects to zero degrees Fahrenheit; a 1:1 mixture protects to 34 below zero.

Building a dry well

A dry well is a rock-filled underground pit into which non-effluent surface water can be drained. It works something like a seepage field. Water filters through the well's porous fill, then eventually soaks into the surrounding soil.

Although other applications are practical, the most common usage for dry wells is the disposal of roof runoff via the gutter and downspout system. If, for aesthetic reasons, you don't want downspouts to drain directly onto your yard, or if rainfall in your area can be extremely heavy and erosion becomes a problem, then a dry well may be the answer. A dry well also may be used to dispose of water from a floor drain, or backwash/recharge from a water softener.

Dry wells can be simple or complex. The simpler ones involve little more than a hole dug into the ground, then filled with rock, gravel, or old bricks and concrete blocks. Almost anything that will not absorb water, yet will leave air pockets through which water can filter, will work as dry-well fill. Fancier wells can be fashioned by chiseling holes through a steel drum, then filling the drum with rock or masonry rubble.

Where the volume of water is so great that a single dry well just won't handle it, a whole series of wells is possible. Provide an outlet near the top of one well and, when the level in it reaches that outlet, water will flow out into another dry well. A series of wells has the effect of spreading the water absorption over a large area and minimizes the possibility of any water resurfacing.

Dig a hole in the ground to form as large a

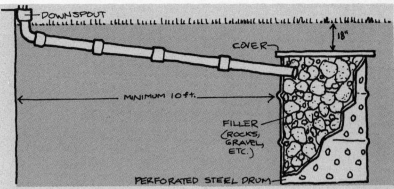

The most practical way to dispose of excess surface water is through a dry well: A steel drum or just a rubble-filled hole in the ground. Use a solid cover to force water to percolate downward and outward, not upward.

dry well as you need. Of course, if you're using a steel drum, the hole needs to be only large enough for it. Regardless of the type of dry well you install, it should be no less than 10 feet from your house foundation wall. This prevents seepage into your basement. And keep the top of the well at least 18 inches below grade.

Downspouts can be connected directly to the dry-well inlet, using 4-inch sewer pipe for the underground portions. (See page 55 for how to work with various types of sewer pipe.) Be sure that the water dumps into the top part of the well, so that it can filter downward through the fill. Inlets should slope about ¼ inch per foot away from the bottom of the downspout, and should be placed below frostline if you want them to work year-round. Where the frostline is deep, a dry well may have to be installed more than 18 inches below grade.

What it takes

Approximate time: Digging for the dry well and digging a trench for its inlet take the most time and are the most work. From there, you're limited only by the time it takes to chunk in the fill, make the necessary pipe connections, and backfill with soil.

Tools and materials: No huge collection of tools is needed for this job. You'll need a sharp-bladed shovel, of course. Breaking up excessively large rocks or blocks for the fill material may be done with a sledge hammer, but be sure to wear eye protection. The use of PVC drain pipes is recommended, and they require PVC solvent-welding cement and an applicator brush. Other drain pipes, using a different connection method, may require other special jointing methods (see page 51).

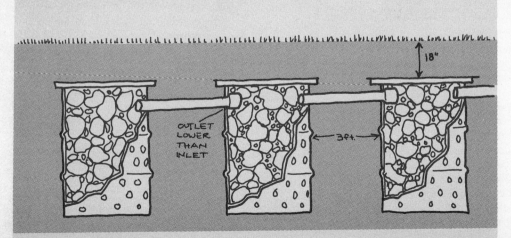

Hooked in tandem, a series of dry wells can dispose of large amounts of water. What the first well can't handle passes to the next, and so on until all the water is absorbed. Outlets should be lower than inlets to prevent back-up.

PROJECTS 91

Glossary

Air chamber. Prevents water-hammer in a supply system by providing an air cushion for fast-flowing water to bounce off when flow is stopped.

Appliance. General term applied to any water-using device that is somewhat mechanical in nature.

Aquastat. Device in a hot-water heating system that turns the heat producer on and off to maintain circulating water at a preset temperature.

Bleeding. As applied to heating systems, refers to opening small valves on convectors or radiators to release air trapped in water or steam lines.

Branch drain. Extends from fixtures and appliances to the main building drain.

Check-valve. A valve that allows fluid or gas to flow in one direction only.

Circulator. A pump driven by an electric motor. In forced-hot-water heating systems, the circulator pumps water from the boiler to the radiators.

Cold-water main. Brings cold water to the vicinity of the fixtures and appliances using it.

Collector. A roof-mounted device in solar heating systems. The collector traps heat from the sun and transfers it to a circulating liquid.

Condensation. Water released by warm air when it comes in contact with a cooler surface.

Conduction. Method of heat transfer in which heat moves through solid material toward cooler air or liquid.

Convection. Heat movement characterized by rising warm air and falling cold air.

Cross-connection. A link between the potable water system and a nonpotable, possibly contaminated, water source.

Damper. Device for controlling the flow of air through a duct or flue.

Dielectric coupling. Used to connect pipes of unlike metal types, such as copper to galvanized steel. Prevents electrolytic corrosion between the two metals.

Draft control. A flapper vane that controls air flow in an oil-burner stack.

Drain tailpiece. Section of drain pipe extending below a fixture. Connects to the fixture at the top and the fixture trap at the bottom.

Drain-waste-vent system. The final portion of a plumbing system. Collects wastes, routes them to the sewer, and vents gases to the outside. Often abbreviated "DWV."

Effluent. As applied to plumbing, this is the liquid waste material discharged into sewer system.

Expansion tank. A tank in a closed-loop hot-water system that allows for the expansion of water when heated.

Fan-and-limit switch. A device that controls blower operation in a hot-air system.

Faucet tailpiece. Lengths of pipe extending below a faucet used to connect it to the water-supply system.

Fixture. General term applied to water-using devices such as sinks, lavatories, and toilets.

Fixture-shutoff valve. Allows selective shutoff of individual fixtures and appliances without affecting the rest of the water-supply system.

Hammering. Characteristic noise in water and steam systems caused by rising water or steam meeting cool air or water trapped in pipes or radiators.

Heat control. That portion of a heating system that controls the heat producer to maintain a desired temperature. Includes thermostats and aquastats.

Heat distributor. The air, water, or steam that carries heat from the heat exchanger to the living areas of the home.

Heat exchanger. Part of the heating system where heat from the heat producer is transferred to air, water, or steam.

Heat producer. The device that consumes fuel (oil, gas, coal, wood) to produce heat.

Hot-water main. Begins at the water heater and carries hot water to the vicinity of the fixtures and appliances.

Humidifier. A device for adding moisture to the air.

Humidity. The moisture content of air. The warmer the air, the more moisture it can hold.

Hydronic. Applied to any heating system using water as a heat distributor. Frequently used to describe electric hot-water systems.

House sewer. Connects with the main building drain to carry wastes to a municipal sewer or private septic system.

Leach field. See *seepage field*.
Local code. Set of construction regulations administered by a local government agency.

Main building drain. Carries house wastes outside the house foundation, where it connects to the house sewer.
Main shutoff valve. Valve that cuts off the entire house water supply.
Main stack. One or more vent stacks that can serve more than one fixture or appliance, including a toilet.

National plumbing code. Nationally recognized code that forms the basis for many local plumbing codes.

P-trap. A type of fixture trap designed for a waste pipe located in the wall.
Pilot jet. Small flame in a gas furnace (or gas range) that ignites the main burner.
Plenum. A large air chamber to which individual air ducts are connected in hot-air heating systems.
Plumbing. The pipes, fixtures, and appliances comprising a building's water-supply and drain-waste-vent systems.

Radiation. The direct transfer of heat from a warm object to a cooler one.
Revent. A pipe installed to carry gases only. It connects to a main vent above waste-water lines.
Riser tube. Small, flexible piping that eases the connection between the water-supply system and a water-using device. Eliminates the need to line up a faucet tailpiece exactly with its water-supply pipe.

S-trap. A type of fixture trap designed for a waste pipe located beneath the floor.
Secondary stack. A vent stack, usually of smaller diameter than a main stack, that serves fixtures other than toilets.
Seepage field. An area where liquid effluent from a septic tank is allowed to percolate into the soil.
Septic tank. A tank, often concrete, that uses organic action to break down wastes. Called for where a public sewer system is not available.
Service entrance. Supply pipe bringing water into a building. Usually located with water meter and main shutoff valve.
Sewage-treatment plant. A more effective private or public sewage-disposal system than a septic tank.
Soil stack. Another name for a main stack serving a toilet.
Stack relay. A safety device used on oil burners to turn off the flow of oil if ignition does not occur.
Stop-and-drain valve. Allows water to be drained from the water-supply system to prevent freeze-ups. Also called "stop-and-waste."

Temperature-and-pressure-relief valve. The safety valve on top of a water heater that bleeds off potentially explosive temperatures and pressures in case of runaway heater operation. Abbreviated to "T&P valve."
Thermocouple. A device that generates an electric current when heated.
Thermostat. A device that controls heating to maintain a nearly constant temperature.
Trap. The P- or S-shaped device in a fixture drain that holds a small amount of drain water, preventing the entry of sewer gases and vermin into the building.

Vapor barrier. A reflective surface on insulation which faces the warmer (inside) air. The barrier prevents condensation from occurring inside building walls.
Vent stack. Piping that reaches up through the roof of a building to vent sewer gases to the outside. Can be either a main stack or a secondary stack.

Water-supply system. The first part of a plumbing system. Brings both hot and cold water into the house and distributes it to the fixtures and appliances.
Well. An underground, private water source that usually uses a pump to extract water for use.
Wet vent. A pipe that serves as both a drain and a vent. Practical when the distance between the trap and vent is short and local codes permit.

Zoned heating. Heating systems (usually hot water) having two (or more) independent circulating systems.

Index

Aerators, fixing, 20
air chamber, water-supply system and, 6, 7
angle-stop shutoff valve, 23
appliances, water
   hookups, 10-11
   repairing, 58-65
      clothes washer, 58
      dehumidifier, 62
      dishwasher, 59
      sump pump, 60-61
      water heater, 63
      water treatment appliances, 64-65
   See also names of appliances

Basement floor drains, freeing, 37
baths/bathtubs. *See* sinks, baths, lavatories
bleeding heating-system line, 81
branches, plumbing, 6
   drains in, freeing, 38-39
building drain, in DWV system, 8

Cast-iron DWV piping, 49
caulking, sink or bath, 29
chisel, 13
cleanouts, in DWV system, 8
clogging. *See* drains
clothes washers
   hookup, 10, 11
   troubleshooting, 58
cold-water main, in water-supply system, 6
cold-weather house shutdown, 90
condensation, curing, 42
conduits, in heating system, 80
copper, sweat-soldered
   in DWV system, 52
   joining tubing, 47

Dehumidifiers, 62
   maintaining, 62
disc faucet, repairing, 19
dishwashers
   connections, 10, 11
   repairing, 59
distributors, heating system and, 80
diverter valve, cleaning, 21
draft control, oil burners and, 77
draining heating system, 81
drains, unclogging, 36-39
   branch and main drains, 38-39
   frequent clogging, solving, 39
   floor drains, 37
   sink and lavoratory drains, 36-37
   toilet drains, 38
   tools for, 15
   tub drains, 37

drain-waste-vent (DWV) system, 4, 5, 8-9
   piping
      cast iron, 49
      copper, sweat-soldered, 52
      plastic, 50-51
      running pipes, 53-55
dry well, building, 91

Electric heating, 79
exchangers, heating system and, 80
expansion tank, draining, 81

Faucets, 16-23
   installing, 24
   lever-type washerless faucets, 21
   repairing, 16
   replacing, 25
   valves for, 22-23
   washerless faucets, 19-21
   washer-type faucets, 16, 18-19
faucet spanner, 15
fire extinguisher, 13
fireplaces, 85
fixtures, plumbing
   hookup, 10-11
   modernizing, 24-25
   See also appliances; faucets; sinks, baths, lavatories; toilets
fixture-shutoff valves, 6, 7, 22, 23
   installing, 28
fixture-waste pipe, in DWV system, 8
flaring tools, 14
flush toilets, 30-32
   mechanisms, 30
   troubleshooting, 31-32
flushing and refilling heating system, 81
   See also tankless flush toilets
forced hot-air heating system, 82-83
   adjustment and maintenance, 82
   blower speed adjustment, 83
   pulley alignment, 83

Garbage-disposer connections, 10, 11
gas burners, 78
globe valve, 22
gloves, work, 13
goggles, 13, 15
ground-key valve, 22-23

Hammers, 12
hand-dug well, 69
heating, 74-89
   cold-weather house shutdown and, 90
   electric heating, 79

   elements of system, 75
   exchangers, distributors, and conduits, 80
   fireplaces, 83
   forced hot-air systems, 82-83
   gas burners, 78
   glossary, 92-93
   heat producers, 76-77
   heat pumps, 88-89
   humidity, 75
   maintenance, routine, 81
   movement of heat, 74
   oil burners, 76-77
   solar heating, 86-87
   steam-heating systems, 84
   See also water heaters
heat pumps, 88-89
   efficiency, 89
   functioning, 89
hot-water main, water-supply system and, 6
house-service entrance pipe, water-supply system and, 6
humidity, 75

Jet pumps, 69

Lavatories. *See* sinks, baths, lavatories
leaks, plumbing system
   fixing, 40
   troubleshooting, 41
      faucet leaks, 20
levels, 13
lever-type faucets
   repairing, 19
   washerless, 21
low flow, problems with, 57

Main drains, freeing, 38-39
main-shutoff valve, water-supply system and, 6
main stack, in DWV system, 8
measuring tape, 13

National Plumbing Code, 5

Oil burners, 76-77
   draft control, 77
   maintaining, 77
   restarting, 77
   stack relay, 76

Percolation table, 71
permits, for plumbing work, 5
piping, 44-57
   adapting pipes, 56-57
   cast-iron DWV piping, 49
   cutters for, 14
   freezing, preventing, 42
   joining sweat-soldered copper tubing, 47
   low flow in, problems with, 57
   plastic DWV piping, 50-51
   running pipes
      DWV pipes, 53-55
      water-supply pipes, 48
   sweat-soldered copper DWV pipes, 52
   thawing pipes, 42
   threaded water-supply piping, 46
   vinyl water-supply piping, 44-45
plastic piping
   DWV piping, 50-51
   tools for, 15
   See also vinyl piping
pliers, 12-13
plumbing system
   basic features, 4-5
   cold-weather house shutdown and, 90
   fixture and appliance hookup, 10-11
   glossary, 92-93
   home-buyer's checklist, 5
   problem solving, 36-43
      condensation, curing, 42
      drains, unclogging, 36-39
      freezing, preventing, 42
      leaks, fixing, 40-41
      thawing pipes, 42
      water hammer, curing, 43
   tools for, 12-13
   See also drain-waste-vent (DWV) system; water-supply system; water systems, private; *names of appliances and fixtures using water (e.g.,* dishwashers; faucets; toilets)
pot-type oil burner, 76
pressure-reducing valve, 23
pressure tanks, 67
pressure-type oil burner, 76
propane torches, 13
pumps, plumbing system
   connecting pump, 67
   jet pumps, 69
   lowering submersible pump, 68
   for residential wells, 66
   troubleshooting well pumps, 68
   See also heat pumps

Residential wells, pumps for, 66
revent, in DWV system, 8

Saws, 13
screwdrivers, 12
seepage field, building, 72
septic system, 70
   installing septic tank, 71
   tank-size table, 71
sewage treatment, private, 70-73
   building seepage field, 72
   building sewer, 71
   installing septic tank, 71
   septic system, 70
   sewage-treatment plants, 73
      for single family, 73
sewer, building, 71
   sewer pipe, 55
showers
   freeing drains, 37
   upgrading shower head, 27
shutdown, of house in cold weather, 90
sinks, baths, lavatories, 24-29
   caulking, 29
   freeing drains, 36-37
   installing faucets, 24
   installing traps, 26-27
   modernizing fixtures, 24-25
   replacing sink sprayer, 27
   upgrading shower head, 27
solar heating, 86-87
soldering pipes, 15
solvent welding, 51
sprayer, sink, replacing, 27
stack, in DWV system, 8
stack relay, oil burner, 76
steam-heating systems, 84
stop-and-waste valve, 22
straight-stop shutoff valve, 23
submersible pump, lowering, 18
sump pumps, 60-61
   connections, 11
   replacing, 61
sweat-soldered copper
   DWV, 52
   joining tubing, 47

Tankless flush toilets, 33
   troubleshooting, 33
   valves, types and parts, 33
temperature-and-pressure relief valve, 6
thawing pipes, 15
threaded water-supply piping, 46
toilets, 30-35
   cold-weather house shutdown and, 90
   flush toilet mechanisms, 30
   freeing drains, 38
   replacing toilets, 34-35
   tankless flush valves, 33

tanks, troubleshooting, 31-32
tools, plumbing, 12-15
   how to use, 12-13
   for plastic pipe, 15
   tips for, 14-15
traps
   adapters, using, 10
   in DWV system, 8
   installing, 26-27
tubes/tubing. *See* piping
tubs. *See* sinks, baths, lavatories

Vacuum-breaker valves, 23
valves, plumbing, 22-23. *See also names of valves (e.g.,* fixture-shutoff valve)
valve seats, fixing, 18
vents, in DWV system, 8
vent stack, running, 51
vertical well systems, 66
vinyl water-supply piping, 44-45

Washerless faucets, 19-21
   lever-type, 21
washers. *See* clothes washers; dishwashers
washer-type faucets, 16, 18-19
   replacing washers, 17
water hammer, curing, 43
water heaters
   caring for, 63
   connections, 10, 11
water softeners, connections, 10, 11, 64
water-supply system, 4, 6-7
   piping
      running, 48
      threaded, 46
      vinyl, 44-45
   private system, 66-69
water systems, private, 66-73
   sewage treatment, 70-73
   water supply, 66-69
water-treatment appliances, repairing, 64-65. *See also* water softeners
welding, solvent, 51
wells
   building dry well, 91
   drilling well, 67
   hand-dug well, 69
   pumps for residential wells, 66
   types of wells compared, 69
   vertical well systems, 66
   See also pumps
wrenches, 13, 14, 15